The Education of a Mathematician

Updating the Ghost of Mr. Jefferson.

The Education of a Mathematician

Philip J. Davis

A K Peters
Natick, Massachusetts

Editorial, Sales, and Customer Service Office

A K Peters, Ltd.
63 South Avenue
Natick, MA 01760

Library of Congress Catalog Card Data

Davis, Philip J., 1923–
    The education of a mathematician / Philip J. Davis.
        p. cm.
    ISBN 1-56881-116-0 (alk. paper)
    1. Davis, Philip J., 1923– 2. Mathematicians–United States–Biography. I. Title.

QA29.D38 A3 2000
510'.92–dc21
[B]   00-020739

Frontispiece and back cover illustrations by Marguerite Dorian.

Front cover photos by Carolyn Artin.

Cover design by Marshall Henrichs.

Printed in Canada
04 03 02 01 00   10 9 8 7 6 5 4 3 2 1

To Bernhelm and Sussi Booss-Bavnbek

For inspiration, help, and friendship.

Table of Contents

Prologue: The Impetus

I had just given a lecture at Roskilde University, Denmark, on the relationship between mathematics and common sense. During question time, Professor Mogens Niss, who was in the audience, asked me how over the years, I had come to my views. I answered that I had never really thought about the question. I assured him, however, I had not received a message as did Moses from the Burning Bush, I had trod no road to Damascus with St. Paul, no visions or revelations emerged from an overheated porcelain stove for me as for Descartes, no apple bumped my head as it did Newton's, my personal goddess imparted no inside information to me as she did for the famous mathematician Ramanujan. It was just a slow and steady progression of day-to-day experiences that added up to a set of opinions.

"Well, why don't you write them up?" my questioner urged. Other people in the audience spoke up and seconded his suggestion. In point of fact, I already had a writing project on my plate, jointly with Prof. Bernhelm Booss-Bavnbek of the same faculty, which in some measure overlapped with what had just been suggested. The idea was that we would bring the Ghost of Thomas Jefferson to date on the current state of mathematics and how we thought it connected with society.

Why Jefferson of all people? Well, despite the contradictions in Jefferson's life and the current irrelevance of some aspects of his political idealism, Bernhelm and I both admired him greatly. And there was more than this: Jefferson was a man who knew a bit of mathematics and more than a bit of science. He knew the leading French mathematicians and scientists of his day, and he said that if he had not gotten involved in politics, he would have devoted his life to these things. His lifelong yearning to be involved with more science and mathematics has been ignored by most of his biographers. So why not use

Jefferson as a device to put forward some of our views on the role of mathematics in today's civilization?

As far as a briefing of Jefferson's Ghost on recent developments is concerned, the problem is that there has been so much new and significant mathematics developed in the past two centuries, so many new mathematizations introduced into our everyday lives, that it is beyond any one person's ability to summarize them all, let alone to understand them. It would be difficult if not impossible to make a list, agreed to by all authorities, of what the ten major conceptual mathematical advances since Jefferson's day have been.

Nonetheless, if Mr. Jefferson would like a very quick review of developments in mathematics, I would say that I, in my professional career, embody that update; that I and ten thousand or perhaps a hundred thousand other mathematicians all over the world embody the update. Our jobs, our knowledge, the things we do with our mathematics did not exist in his day.

It's very easy to get ideas and make plans; to carry them out, to insert them into the larger rhythm of one's life is another matter. I put "Jefferson" on the back burner in favor of previously formulated plans. Bernhelm, for his part, found that he was committed to work with a colleague on *Elliptic Boundary Problems for Dirac Operators*, a deep mathematical monograph.

Months passed during which "Jefferson" lay dormant. The celebrations marking the 250th anniversary of his birth were over. Bernhelm was engaged in a sequel to *Dirac Operators*. On the other hand, reacting to Mogens Niss's suggestion, I spotted a window of opportunity for myself. I talked the matter over with Bernhelm. He gave his generous *nihil obstat*, and I declared my independence from our old joint project while keeping the thought of informing Mr. Jefferson in some manner.

So here goes. I hope to use my own education and professional life as a framework for discussing a few recent developments in mathematics that I think Mr. Jefferson would be interested to know. In that sense, what I have written here is a *Bildungsroman* as well as a series of reflections on the interconnections among mathematics, science, and philosophy.

In my telling, time will occasionally occur non-sequentially. A day-by-day presentation would become both tiresome and chaotic. There may be an occasional mild fabrication where memory has grown dim. The names of some persons have been replaced by pseudonyms.

A final remark to the readers: there is very little technical mathematics in this book, and what there is can easily be skipped over without loss of the principal ideas.

The novelist ceases to be an artist the moment he bends his criticism to the exigencies of a thesis; but he would equally cease to be one should he draw the acts he describes without regard to their moral significance.

—Edith Wharton, *Fiction and Criticism*.

Part I

In the Beginning

Lend Me Your Ears

I was born with an academic spoon in my mouth. I was the youngest, by far, of four children. By the time I was four, my oldest brother was a freshman at MIT, and my other brother and my sister were in high school. Some of my first memories were of them learning Shakespeare by heart for their English classes by repeating the passages out loud. "Friends, Romans, Countrymen, lend me your ears." I could recite them all parrot-like, long before I had any inkling of what the words meant.

From my older brother I learned that there were positive numbers and negative numbers. I learned that there was something called x and I had no trouble at all comprehending what x was. The number or symbol x was an amphibian. Sometimes it was the mystery number that had to be unmasked and identified with an ordinary number. Sometimes it was just plain x, and all one had to do with it were sensible things like x plus x equals two x. Was this an instance of Formalism?

I've heard from some non-mathematical contemporaries that x was an absolute stumbling block in their tottering steps up the slippery slopes of the mathematical Parnassus. Have you ever read *The Vexations of A.J. Wentworth B.A.*, by H. F. Ellis? It's the greatest story about mathematical education ever written. "Let x equal an egg," said one of Wentworth's scholars.

I've also heard that there are deep philosophical discussions as to what the unknown x and the variable x are and what they aren't. To comprehend x perfectly, so it is said by the sophisticated, one must go through three phases: the rhetorical, the syncopated, and the reified. But to me, at the age of six or seven, x posed no difficulty whatever. Children reify quite easily. x might even be an egg. Philosophers complicate matters enormously. That's their métier.

Phil Davis age 3, 1926.

From my early experience then, I recommend strongly that everyone be born with older brothers and sisters. There are minuses, of course, that come with the position of the youngest in the family. Nothing is free.

Not only were my brothers and sister steeped in academics, so also were my cousins and their circle of friends. Is college cost-effective? This would have been an absolutely stupid question in 1930. The main problem then was whether a family could afford to send its children even to a state university or to a local "normal school."

Cousin H.

I have given elsewhere several independent accounts of how I became a mathematician. Every single one of them is right. There may be overlap, but there is no contradiction. I am about to give another account. I became a mathematician because my grandfather raised pigeons. This kind of statement is known in physics as causality. Listen, and you shall hear.

There was a stable behind my grandfather's house. I don't think he ever owned a horse, but he garaged his brand new 1922 Star there. The Star had glass windows. Think of that! Our own car had celluloid flaps.

There was a dovecote just under the stable roof, and it was alive with families of pigeons. They came with the house when my grandfather bought it, and he had a soft spot for them. They were good luck to him, like the storks in Europe, I suppose. He was not interested in racing them, just in feeding them. Periodically he would wander over to Fred Barlow's hay, grain, wood, coal, and ice yard and order a couple of sacks of feed for his pigeons.

Barlow's yard was one of those places in my home town where men would sit around the pot-bellied stove and gossip. The date was June, 1930.

"Mr. Davis, could you use a sweet little race horse?"

My grandfather could never let an opportunity pass.

"What've you got, Mr. Barlow?"

"I've got a little number called 'Sunday's Sweetheart.' She's full of moxie, Mr. Davis. She's a sure fire thing."

"How much you want, Mr. Barlow?"

"One hundred dollars, Mr. Davis, and I'll throw in all the leather."

(A daily newspaper at that time cost 2 cents and tuition at the better private colleges was $300.)

"Well, Mr. Barlow, I'll let you know."

At this point, I'd better interrupt the conversation and tell my readers two things. First: Fred Barlow ran a livery stable in his yard and had a number of horses to rent out. Don't think that Hertz or Avis was the first in the business of leasing transportation. For all one knows, when Cleopatra barged up the Nile with Anthony aboard, she rented the barge and the musicians for the occasion.

Second: to a business man such as my grandfather, all commercial propositions were equally valid: at different times in his life he had been an innkeeper, a tailor, a tobacconist, a soap manufacturer, a soda-pop bottler (it was called tonic in my part of the country), and a distiller of hard spirits. The Eighteenth Amendment to the United States Constitution wiped out his distillery, and except for that one small fact, I would now be a very rich man running my own foundation or establishing chairs for professors to sit in and practice academic insufferability. This wipe-out is now considered an instance of man-made catastrophe theory.

To return: this is the point where Cousin H. entered the picture. Cousin H.–that's what he was called by one and all–had just graduated from MIT, majoring in chemical engineering, and June, 1930, was a very bad time for anyone to graduate, irrespective of his major. Cousin H. had no job that summer and his prospects were dim.

My grandfather asked Cousin H. whether he would like to race a horse at Rockingham Park four miles away in Salem, New Hampshire. If so, he would go 50-50 with him and even lend him his share of the up-front money. Cousin H. replied that all he knew about racing was what he'd picked up as a kid selling papers at Rockingham, but he had a friend George, also unemployed, also an MIT graduate, who knew rather more. He would put the proposition to George. Cousin H. was being modest. His experience in selling papers and hanging around the Rockingham paddock had gained him much in the way of racing smarts. You will see.

George, it turned out, was willing. And so a three-way partnership was concluded, with my grandfather owning 50%, Cousin H. 25% and George 25% of a filly named Sunday's Sweetheart, stabled momentarily in Barlow's yard and whose parameters were as unknown as the x's in a mathematical equation.

And so Cousin H. and George were thrown immediately and brutally into the world of horseracing, with its need for trainers, equipment, feed, stable and entrance fees, with its jockeys, its handicapping and bookmaking, all accompanied by manipulations that were well known but mentioned only in hushed tones around the pot-bellied stoves. And this is where I, at the age of seven, came into the picture.

The partnership was one week in operation, when Cousin H. asked me whether I would like to act as bookkeeper. Here is how he explained the duties involved.

"I'll get you a notebook. In one part you write down all the expenses. Like 50 cents for hoof salve. In the other part you write down Sunday's Sweetheart's earnings. Like The White Mountain Handicap: $200. Assuming she wins it. And then, from time to time, you've got to add up these things and figure out the three way split: 50%, 25%, 25%. Think you can do all that?"

I was flattered. After all, Cousin H. was my favorite cousin. I agreed to be bookkeeper. At the age of seven my addition was adequate; my division might have to be bolstered by adult techniques. Of course, it never came to that.

Cousin H. brought me a green-stained but mostly empty lab notebook from his course on quantitative chemical analysis. I printed "Sunday's Sweetheart" on the cover, and entered the first item: hoof salve 50 cents.

My duties as bookkeeper turned out to be remarkably light, short, and educational. Sunday's Sweetheart proved to be rather deficient in the moxie promised by Fred Barlow. Cousin H. and his partner George ran their Sweetheart twice and, poor thing, she didn't come anywhere near the money. The partnership decided to allow Fred Barlow to repossess the filly at 25% of the selling price, and then to dissolve itself.

I was heartbroken. I was pulling for the Sweetheart all the way. But Cousin H. took me aside and gave me my first lesson in what he called the realities. The facts of life.

"Now Phil, you have to understand that things are always more complicated than they seem. You might think that the reason for owning a racehorse is to run the horse and to win a purse. That's only part of the story. The other reason is to be able to gain information denied to the general public. You mix with the other owners and the other jockeys and you hear things. An owner can bet on other horses. And it hasn't been too bad for me. George and I made up our losses and a bit over."

Having said that, he paid me 10 cents for my services as bookkeeper and gave me the lab book as a bonus.

The empty lab notebook played a role in my subsequent career as a mathematician. When I was eleven or twelve I wrote down in the lab book "secret things" about numbers that I had found out for myself. Things such as the sum of two odd numbers is always even. The product of three consecutive numbers is always divisible by six. By the time I went to college, Cousin H.'s lab book contained what is known in mathematics as the Newton formula for polynomial interpolation, a substantial but by no means remarkable achievement.

Of course, it took me some years before I understood what really went on at the track. Cousin H.'s motto was "Inside information beats the stats ten to one." That's why it's illegal.

As I grew older he would repeat the motto over and over to me. It was his mantra. Sometimes he put it this way: "If you read about it in the papers, it's too late." It was my first lesson in probability theory and on the relationship between a deterministic world and a probabilistic world, a problem that bugged Einstein and continues to bug post-Einsteinian science. God has inside information, Einstein thought. If we ask Him in the right way, He will tell us. That is one of the basic philosophies of physics.

Cousin H., particularly after he had retired, became a habitué of the track, a lover of the gee-gees, as they used to say. I would ask him: how can you go up to the windows and plunk down your money when you know full well that horse racing is full of sin and corruption? How can you find a rational basis for placing a bet? And he would answer: sin and corruption are certainly there. They are part of the world as we experience it. However: inside information conquers all. But too much inside information can lead to divine or civil punishment. Ask Prometheus to tell you his story.

In the years that followed, information, its gathering, its dissemination, its use, its misuse became the touchstone of the computer age. Information was created, destroyed, faked, transformed, filtered, sought, bought, sold, transported, restricted, interpreted, misinterpreted, theorized. The word "database" became part of the vocabulary of kids in grades K through sixteen. The relation of information to knowledge and to wisdom were the subjects of frequent polemical essays. Who shall have access to information and who shall not, who shall pay for it and who shall not, became one of our major ethical problems.

Hitting It Big

Let me return to Rockingham Race Track. It was not all fun and games and equine bloodlines with a few dirty little tricks thrown in here and there. The track (and all gambling schemes) have ruined lives. I've known this since childhood.

I used to go along with my father and mother to visit a prosperous farm in southern New Hampshire owned by immigrants from Poland. We would come back with fruit (mostly blueberries), vegetables, chickens, turkeys, raw milk (for hanging up over the kitchen sink to make pot cheese). One of the farmer's sons went to Rockingham and became addicted; then the farmer himself. First the farm was mortgaged. Then it was sold at auction. Multiply this case by a million.

Gambling, a human activity that has existed since recorded history, is now very big business in this country. And it costs the country billions in damaged lives. Some of the sting has been taken out linguistically by calling gambling "gaming." This puts the activity in the entertainment category. The right word in the right place can work wonders. This is called semantics.

A recent letter in a local newspaper written by a high school teacher of mathematics complained that the math teachers of the country have not done their duty as regards gambling. They have not driven home the brute fact that every time one goes into a casino or plays the lot, on average, one loses money.

I have a somewhat different opinion. People know very well that, on average, they lose money. They know that the casino owners or the state governments are raking it in; otherwise the slots and the wheels and the little scratchaway slips would be junked rapidly and the value of the stock of the big suppliers of mathematical gambling systems and hardware would plummet. It is not a question of knowledge; there is something else at work.

Mary Y. and John Q. Public seem to say to themselves: "Averages be damned! I'm not going to run my life by averages. I may really hit it big next time." And if they hit it, by God, fifty million Powerball dollars will walk into their checking account over a period of years, to make them ecstatically happy or to make them miserable, as the case may be.

What is the relation between the accepted mathematical theories of probability and people's actions? Is the former an adequate predictor of the latter? Is it possible to set a dollar value on the entertainment implicit in gambling, on the expectation of "hitting it big" and to factor the adrenalin highs into our theories? Well, you might say that the market does just that automatically, and its answer seems to have resulted in the largest mania for gambling and the largest legalized gambling industry that has ever flourished in this country. At the same time, a small-time operator in Woonsocket, Rhode Island can be hauled in for peddling football lottery tickets.

Speaking of adrenalin, the Marquise de Chatelet (1706–1749), a mathematician and an inveterate gambler, about whom I will have rather more to say later on, remarked in a book she wrote on how to live right (*Reflections sur le Bonheur*), that the emotions of the gambler run the gamut from hope to fear. This, she said, is a very good thing since it serves to keep the gambler in good health. In her day there was probably no pamphlet entitled *Early Signs of Compulsive Gambling* distributed by Gamblers Anonymous that analyzed the Chatelet syndrome and numerous related compulsions.

Beyond gambling conceived narrowly as The Lot, The Wheel, or the lottery tickets that kids sell to enable their high school band to buy uniforms, we are all fundamentally gamblers. When we cross the street we take a chance, when we have children we take genetic chances. Pop-philosophers assure us that life is a worthwhile gamble and so we indoctrinate our children into our personal perception of the common sense of living in a chancy world. And what role does mathematics play in all this?

I recall going to a scientific meeting some years ago at the National Cancer Institute in Bethesda, Maryland. A noticeable number of the participants, including the local statisticians, were puffing away. I wondered how this could be since the anti-smoking campaign had intensified considerably. Apparently statisticians don't let their public conclusions determine their private behavior.

Let's talk about travel. A large fraction of the population travels. There we may experience the reverse phenomenon in which a single disaster sets off a chain reaction in excess of what might have been deduced from a calm consideration of the probabilities. About ten years ago, in response to terrorist activity in London, the tourist business experienced a terrible slump. So much so, that when my wife and I showed up in London as we had planned months in advance, our British friends expressed surprise: "You didn't cancel your trip?" Or, please tell me what is the prudent course of action to take when you read in the papers that a certain airline has had a disaster and you are holding space on one of their flights.

Yes, one day we say "Damn the stats. The average weather in an average life is a light drizzle. All honor to the individual life with its sunshine and downpours!" And the next day we succumb to the insurance peddler who points out some corner of our lives that is not covered by our current policies or by the government. Reader: does your homeowner's policy cover you against power blackouts and spoilage of the contents of your freezer? Check the fine print.

Conclusion: it is very hard to move from mathematical theories of probability to moral or economic advice to the individual. Whether or not such advice will be taken or whether it would even be reasonable or sensible to take it seems itself to be a random variable.

Rational Recreations

There's a folk saying in my profession that mathematicians make good musicians. I've never heard it stated the other way around. In my case, it was only partially true. From the age of ten or so, I took to classical music readily. I grasped the symphonies, tone poems, and overtures immediately–the ones that would now be called symphonic chestnuts: Haydn's *Clock Symphony*, Mozart's *G-Minor*, Tschaikovsky's *Fourth*, etc. I knew them by heart. As I played the old 78s on our large wind-up phonograph, I loved to follow the movements on such scores as I had.

I would say that my perception of these classical pieces in those early years was as close to Platonism as I have ever been, mathematics included. I could not conceive that a single note of the *G-Minor Symphony* might have been other than it was. Mozart's classic had pre-existed the creation of the universe.

When it came to playing an instrument, matters were quite different. Piano lessons for two years, then clarinet lessons for one. But my fingers simply would not follow the instructions sent to them by my reading of the musical notation. My true instrumental level turned out to be playing simple sixteenth-century dances on the recorder (block flute).

Learning to finger the scales in the way that my piano teacher Mrs. Reed would have me do, moving my thumb under the other fingers was sheer agony. I was frustrated and bored. My eyes moved upward from the pages of music to the top of the piano. There was a set of books sitting there between bookends that was my sheer delight. I abandoned *Für Elise* and took down the first volume. It was an important book in my life.

It's a book that's rare, but not that rare. It comes in four volumes and you can probably pick up the set for fifteen hundred dollars or so. Its full title is

Rational Recreations: In which the principle of number and natural philosophy are clearly and copiously elucidated by a series of easy, entertaining, instructive experiments. Among which are all those commonly performed with cards.

Author: William Hooper. Date: Around 1770. Place: London. There are a number of editions. The set had been bought by my brother. Why? Who knows? He was not really a bibliophile, but it might have been a phase he went through.

What was in Hooper's volumes? "Number" meant arithmetic. "Natural philosophy" in those days meant physics: mechanics, optics, electricity and magnetism (Ben Franklin's kind of experimentation, often displayed as a parlor game: everyone form a ring and hold hands—not to receive holistic energy or as an act of reverential piety and transcendental fellowship—but the beginnings of shock therapy: to receive a round-robin shock from the newly developed electrostatic machine). Also hydraulics, and, would you believe, pyrotechnics (fireworks) and card tricks.

What attracted me particularly was Volume I, *Arithmetic*. It was full of number games. "Take a number, any number. Double it, add three, reverse the digits. Tell me what you get. Ah then, the number you first picked was such and such." I learned from these number games. I soon saw through them. I invented my own games. Thanks to William Hooper, (*floruit* 1770, and that's all I've been able to find out about him) I perceived that the theory of numbers was a very cool subject. I discovered theorems on my own and marked them down in the lab book Cousin H. had given me.

Why I Didn't Go for Marxism

I must have been born with skeptical and satirical genes, if there are such, and I was aware of it early. Regarding skepticism, the precept "question everything" looks good on a bumper sticker, but if one adhered to it rigorously, one couldn't get on with one's life. Part of the mathematizations installed today consists of rating things on scales. "On a scale of one to ten, where do you stand on the issue of race?"; "On a scale of one to ten, how much do you like Norwich Terriers as opposed to Airedales?" I question such ratings. Yet if someone were to ask me: "On a scale of one to ten, how skeptical do you consider yourself, rating Pyrrho of Elis as ten," I would answer "six."

My brother Barney, eleven years older than I, was handsome, athletic, popular, talented, and a 1933 graduate of Dartmouth College. I adored him. He was my god. Sometime in those deep Depression years he became a communist. With proselytizing zeal, he went after me (ten years old) as a potential convert (from what?). Read John Strachey's book *The Coming Struggle for Power*, he advised me, leaving it around conveniently. Since I was the youngest in the family, lots of books were left around for me to dip into. The math books took hold. The others...well, it depended. I've never been afraid to say "I don't understand what's written in such and such a book." We all ought to do it more often. The kids in the classroom often sit mute in their seats, afraid that if they ask a question they'll be considered dumb. And so do adults, particularly men.

One day in 1935, Barney and I drove into Boston to visit the office of the New England Division of the Communist Party of America. My brother had some business with the Head. Address: just off Washington Street in downtown Boston. Downstairs from the N.E.D.C.P. was a delicatessen where wonderful pastrami sandwiches and pickles could be had for fifteen cents. When I think of the American Communist Party, I invariably think of pastrami sandwiches.

Phil Davis' sixth grade photo.

We walked into the head's office. and I remember seeing several mimeograph machines and stacks of pamphlets. There was a large picture of Karl Marx behind the head's desk. I knew who it was. I recall looking around the office trying to find pictures of the head's wife, family, children. I had been in other offices where such things were displayed prominently and proudly. There were none such in this office. Pointing to the picture of Karl Marx, I asked the Head, "Is that your grandfather?"

The Head did not take this question kindly, and replied with something like "Oh, a wiseguy kid, are you?"

This encounter was the first incident on the path to a disconnect. The second came some months later. In the summer of 1936, the national elections were a few months away. Franklin D. Roosevelt was running for his second term as President, Alf Landon was the Republican contender, William Lemke that of the Union Party, and way down the list was Earl Browder for the Communist Party.

The *Sunday Worker*, a paper my brother subscribed to, had a children's section, and it announced a prize for the best poem on the election. The prize was a compass (for drawing circles). I owned plenty of compasses; nonetheless I entered the contest. I won the contest. My entry was printed. I waited for weeks for the compass. The election came. Roosevelt won in a landslide: Browder garnered about 70,000 votes. The compass never arrived. I was irritated beyond measure. My philosophical disconnect with Marxism was now complete.

As for my brother? In those days he was seriously thinking of joining the Abraham Lincoln Brigade to fight for Republican Spain. He died young; not in Spain, but in Mexico. An accident. He did not have to live through the shock of the Stalin-Hitler pact of August 1939.

Much Ado about Nothing

At an early age (but only in certain contexts)—I agreed that it all added up to nothing.

I met my first professional mathematician because my brother Barney was an excellent clarinetist and played in the Dartmouth College Symphony. It was 1934 or 1935 and I was eleven or twelve. Algebra was then entering my life. I had just learned the quadratic formula. I knew that the quadratic equation was $ax^2 + b\,x + c = 0$. I knew what a, b, and c represented. I knew that x was the unknown. I could follow and reproduce from memory the algebraic steps that led from this equation to its solution expressed in the famous quadratic formula

$$x = \frac{-b \pm \sqrt{b^2 - 4ac}}{2a}$$

I knew why there was a plus or minus sign in the formula. But, for the life of me, I couldn't figure out why the left-hand side of the quadratic equation should always add up to zero. I personally knew many quadratic equations in which it did not add up to zero. For example: $x^2 + 3x = 4$.

An aside: In those days, it was considered a standard part of a liberal education to know the quadratic formula. I'm not sure how the formula fares now. Mathematical educators are apt to ask: Does your average dentist have occasion in his practice to use the quadratic formula? Does your average insurance salesman? Will the solution to the quadratic equation bring peace on earth? You're not sure? Then to hell with it.

My reading of educational history reveals that this kind of questioning was advocated by the great Alfred North Whitehead in his famous essay "The Aims of Education," (1922) and has now, in some educational circles, both elementary and university, been parlayed into the proposition that all that is important to know is how to get along with your neighbor. Courses on that topic seem to glut the curriculum, while the quadratic formula is one among many of the endangered species of knowledge.

For months, I nursed my perplexity privately. Ultimately, I opened up to my brother, who by that time had graduated from Dartmouth and was a graduate student in chemistry. I was a big correspondent so I sent my brother a letter and asked him why it should all add up to zero. The back and forth of letter writing took a bit of time. I would never have called him on the phone. In those days people made long distance calls only if there was a death in the family. My brother, well versed in first year calculus from William Fogg Osgood's book, didn't see the point to my question, so I got no immediate relief from that quarter.

That year my brother was rooming in the house of L. L. Silverman, who was professor of mathematics at Dartmouth. Silverman's son Raphael (Dartmouth '36) who later was one of the founding members of the famed Juilliard Quartet, was a friend of his. (He performed under the professional name of Raphael Hillyer.)

A month or so after I wrote him and got his dismissing answer, he wrote again that he would shortly be copping a ride with Professor Silverman who was driving down from Hanover to Boston and that when he came by our house, I could ask the professor my question.

Allow me a moment now to provide a mini-portrait of Louis Lazarus Silverman (1884-1967), who was the first "real" mathematician I ever met. L. L. was short, stocky, and had a twinkle in his eye. At the time, I did not, of course, know L.L.'s mathematical credentials in the professional sense. He was an undergraduate at Harvard and received an MA from Harvard in 1906. He had a doctor's degree, awarded in 1910, from the University of Missouri at Columbia. He worked at General Electric during World War I. I once had occasion to refer to his doctoral thesis "On various definitions of the sum of a divergent series" after I had became a professional. He was at Dartmouth from 1918 to 1953 and on his retirement from Dartmouth, he did some teaching at the University of Houston and at South Texas College. One of his former pupils was the famous Harvard astronomer Harlow Shapley.

Silverman was a man who within his lifetime had moved from a shtetl in Lithuania to a small town in New Hampshire. The little New England villages weren't devoid of interest. They had their poets (Robert Frost, Edward Arlington Robinson), they had their playwrights (Thornton Wilder, Eugene O'Neill) and their short story writers (Celia Thaxter and Sarah Orne Jewett), but on the whole, in that generation, the movement was away from small towns and into New York City. That's where the action was. Read *My Sister Eileen.*

L. L. learned his arithmetic at the age of five from a policeman who came to his parents' inn. He and his parents immigrated to the United States at the turn of the century. In the early Twenties, L. L. made a trip to Russia as did a number of American liberals, and held views about what he saw that were not far from the classic Lincoln Steffen's misappraisal "I have seen the future and it works." As late as 1947, L. L. was sanguine about the progress made in Russia. Perhaps he was struck by the contrast with his memories of his childhood village. My impression of his concerns with political policy in his later years was is that they were of a man who had little feeling for the realities.

What happened in the week before the professor's arrival was that I was able to answer my own question. How stupid I'd been. How simple the answer: one simply moved all the terms to the left side of the equation, and zero was what was left on the right hand side. That was the virtue and the power of negative numbers. The great mathematician Diophantus (third century) did not have negative numbers. Nor did the great Al-Khwarizmi (ninth century), the nominal inventor of algebra. For those ancient fellows, all numbers had to be what we now call positive. They could not have written down $ax^2 + bx + c = 0$, and think of it as the generic case.

The Professor delivered my brother and stopped off at our house for a cup of tea. I was introduced to him and he said to me

"Barney tells me you have a question in mathematics."

"Well, I did, but I think I've figured out the answer."

I explained to L. L. what my question was and what my answer was.

"Yes, you've got it. Mathematicians have a tendency to throw everything onto the left-hand side."

"Why?"

"They think it makes things neat. You know who else did it recently?"

"Who?"

"Einstein. (I'd never heard of Einstein.) Einstein wrote down $G + cT = 0$ and that, he said, sums up the universe. Solve that equation, Phil, and they will carve your name on the front of the physics building."

Professor Silverman drove off to Boston and I was left with the cheerful feeling that the universe, like mathematics, was indeed neat. This feeling evaporated as I got older and began to question whether the universe could ever be summed up either to zero or to something positive or negative in a few lines.

As far as mathematics itself is concerned, its perceived neatness is often described as mathematical esthetics; an ideal constantly to be strived for and occasionally achieved. Let me quote the words of Richard Brauer, a great twentieth-century algebraist. He is talking here about a concept known as mathematical groups, but what he says really has wider application:

> It has been said by E. T. Bell (famous mathematician at California Institute of Technology and writer of detective stories) that "whenever groups disclosed themselves or could be introduced, simplicity crystallized out of comparative chaos." This may often be true, but strangely enough, it does not apply to group theory itself, not even when we restrict ourselves to groups of finite order.... A tremendous effort has been made by mathematicians for more than a century to clear up the chaos in group theory. Still we cannot answer some of the simplest questions.
>
> This may sound as if I am critical of the theory of finite groups. On the contrary, this is why I am fascinated by it. It is the unknown and the mysterious which attracts our attention.

Well, much has been clarified in group theory since E.T. Bell's day, and a good deal of it is hardly neat. And as far as the mathematician's inclination to equate all to zero, I have often wondered whether it is allied to the perception that mathematics is one vast tautology or whether it bolsters the case for the *Creatio ex Nihilo*.

If the reader thinks that zero, zip, cipher, nil, naught, zilch is a simple concept, or just a place holder in the positional notation for numbers, let him or her turn to the pages of Brian Rotman's recent book *Signifying Nothing* and learn the complex semiotics of the famous goose egg symbol.

Lawrence High School and
What I Learned there about Mathematics

Lawrence, Massachusetts, a city thirty miles north of Boston, a few miles from the New Hampshire line and on the Merrimack River, is where I grew up and spent the first sixteen years of my life. The city itself was a latecomer on the New England scene, having been developed in the 1840s by Boston investment capitalists (Abbott Lawrence among them) as a place where textile mills could be located profitably. The river was dammed up, canals built, and water-power tapped. Later, of course, the looms were steam- and electric-driven. When I think visually of Lawrence, even today, I think of the high mill chimneys. When the chimneys were belching forth dark smoke, all was well economically; when they merely sat there against the horizon like dead cylinders of bricks, things were tough. When the mill whistles put out powerful blasts at seven in the morning, they woke me up and I went off to school, but economically, the whistles were ambiguous.

At their high point in the early 1900s, the mills employed 30,000 men, women, and teenagers. One of the first signs of serious trouble came in 1912 with the Great Lawrence Strike, a classic, if one can call it that, in the history of the American labor movement. Serious trouble was anticipated. Civilians were deputized and issued small arms, among them my grandfather, a fact I was not aware of until his revolver was found among his effects after he died in 1934.

Lawrence was hit hard by the Depression of 1929–1939. Mill owners moved their operations south where wages were lower and unions nonexistent. There were strikes in Lawrence and union organizational drives although there was no work. Perhaps my recollection of the Depression was why, a few years ago, I was against the G.A.T.T. (The General Agreement on Tariffs and Trade). Businessmen might worship the Almighty Bottom Line. Leading economists might make their models and conclude "Go for G.A.T.T." And I wondered how they could factor into their equations the human suffering I had seen in the Thirties.

My father, who had owned and operated a very successful men's clothing store, was hit hard, and never really recovered from the Depression years. There were suicides; I grew up with the children of two such men, and this troubled me deeply.

Despite the economic difficulties of the city, despite the fact that there were nasty slum areas where rough lives were often lived, I liked my city and was comfortable with it. I knew all its streets; I knew its history; I knew the many ethnic groups that constituted its population of about 85,000: old New Englanders, Irish, Scottish, French Canadians, Germans, Italians, Jews, Russians, Poles, Lithuanians, Syrians, Armenians. The number of American blacks was numbered at 100, of whom two attended my grammar school. I played quite regularly in the grammar school yard with one.

I never understood why people in the late Thirties looked down on Lawrence; I never felt ashamed of having grown up in a mill town. Even during the Depression, when there was no money, we lived in a nice section of the city, high up on Tower Hill. When I looked out of our windows to the east, I would see the mill chimneys with no smoke emerging; and on a nice day when I looked toward the west, I would see Mount Monadnock on the purple horizon sixty miles off in New Hampshire.

Lawrence High School, located in the downtown area, was a half hour walk. Every day for four years I walked both ways, lugging my books in a leather briefcase. The school was large: 4,000 students, divided into a number of "courses": the college course, the general course, the commercial course, and the shop course for boys and the home arts course for girls. Enrolled in the college course, did I think of myself as part of an elite? Not at all. I was simply following the path of my older brothers and sister. My awareness of the socio-anthropological subtleties of who went to what college and who didn't go at all came at a later date.

And what of the teaching staff at Lawrence High School in those days? Well, I had some men teachers now (unlike grammar school), and the women teachers all had to be either unmarried or widowed. They were college graduates from the fifty or so New England colleges; there were two PhD's among them, and at the other end of the scale, there were a few dum-dums. The rumor or joke or possibly the reality was that in order to get a teaching job in Lawrence, one had to grease the palms of the school committee, a publicly elected group of four.

To some extent students realize their potential independently of their teachers. In my day, it was the boast of Lawrence High School that the famous poet Robert Frost, jointly with his future wife, had been its valedictorians in 1892, and I know that during the decade or so surrounding my attendance, the school sent on a good half dozen who became professors of this and that in the finest American colleges. Two students became concertising instrumentalists on the international scene, a tribute to the high school music teacher, a granitic but popular Vermonter who himself was a graduate of Dartmouth.

And what of my courses? English, Latin, French, German, American history, civics, algebra, geometry, trigonometry, physics, chemistry. A pretty standard college course.

In Latin I got as far as Caesar's *Gallic Wars*; in French, *Colomba* by Merimee; in German, *Immensee* by Theodor Storm: "*An einem Spaetherbstnachmittage ging ein Mann langsam die Strasse hinauf...*" How many hours did I spend— or waste—looking up words in the vocabularies at the backs of these books. In English, one Shakespeare play per year, novels by Walter Scott and George Eliot; a romantic novel called *Queed* by a now forgotten author Henry Sydnor Harrison (1880–1930), which was set in the post-Civil War middle states, and made the point that a rebel southern gal might indeed find happiness in the arms of a Yankee husband. It was a novel of healing and I read it in a day when the last living member of the Grand Army of the Republic in Lawrence tottered down the middle of Essex Street in the Memorial Day Parade.

I didn't mind committing to memory long passages of Shakespeare; many of them I remembered from my brother's and sister's high school days. I feared and was traumatized by having to get up in front of the class and say something I had prepared but was not allowed to read. If I had been told then that I would spend a good fraction of my life speaking in public (I include classroom work), I would have laughed and said, in anticipation of Sam Goldwyn, "Im Possible."

Let me now turn to mathematics in high school. As I have already indicated, I knew the algebra beforehand. Trigonometry was a bit new, but quite algebraic. I loved to derive complicated trigonometrical identities. We drilled endlessly solving triangles with pencil, paper and a table of logarithmic and trigonometric functions, interpolating when necessary for more accuracy. Within ten or fifteen years all this mind-oppressing drudgery would be gone, eliminated by the development of digital computers. The stacks and stacks in libraries of numerical tables of special functions, computed laboriously almost until the mid Fifties, could be hauled off to the dump, with a few copies saved for deposit in some Museum of the History of Mathematics. Do you suppose that restored Williamsburg in Virginia or restored Sturbridge Village in Massachusetts could be talked into taking a few copies?

Plane geometry à la Euclid was genuinely new to me. I loved it immediately. I loved the theorem proving portion of the textbook, and I loved to work the theorem proving problems that were set (which we called "originals"), discovering a logical path from what I already knew to what I wanted to demonstrate. If there was any one thing that hooked me on mathematics (and I've heard the same from a number of professionals) it was the approach to geometry of Euclid and the Greeks. To explain this phenomenon, I like to quote from the short biography of Thomas Hobbes in John Aubrey's *Brief Lives*:

> He [Hobbes] was 40 yeares old before he looked on Geometry; which happened accidentally. Being in a Gentleman's Library, Euclid's "Elements" lay open, and 'twas the *47 El. libri I.*[The Theorem of Pythagoras]. By G—, sayd he (he would now and then sweare an emphaticall Oath by way of emphasis) *this is impossible*! So he reads the Demonstration of it, which referred him back to such a Proposition; which proposition he read, That referred him back to another which he also read. *Et sic deinceps.* [And thus, one after another] that at last he was demonstratively convinced of that trueth. This made him in love with Geometry.

I felt much the same. I have written a long description in *Mathematical Encounters of the Second Kind* of how a certain original called Napoleon's Theorem kept me in thrall for years.

Geometry was deduction, proof. Algebra, on the contrary, (and at least in its elementary parts) was mere computation. Proof represented absolute, rock-bottom certainty; indubitability. Proof in geometry established the fact that some certainty was possible in this unstable world. Proof was conceptual, and geometrical theorems were construed in a dynamic and material sense. If you do this and this and this with a ruler and a compass, then something else will unquestionably happen. It struck me as eerie how a pure mental act could compel the physical universe to behave in a certain way and without physical intervention. In the beginning was the thought, or was the thought simply one way of organizing the beginning?

And then certainty went out through the door as doubt crept in the window. Shyly at first, tentatively. It came about in this way. In an appendix to one of my geometry books, a paradoxical situation was set forth: a "proof" that every triangle is isosceles, i.e., every triangle has two sides that are of equal length. This was palpable nonsense. If one draws a triangle "at random," it very likely has three sides of unequal length.

What about the proof? Starting from a triangle that was not isosceles, one made a few standard constructions, (auxiliary lines they were called), and arguing logically on the basis of the figure that was then created, one arrived at the startling conclusion that two sides of the triangle were indeed equal.

How could this be? I went over the steps of the "proof" a number of times. They all seemed to be correct. Something had to give: if the figure with all its auxiliary lines could be drawn as claimed, then indeed every triangle was isosceles. Conclusion: there was something fishy about the figure. But what? Moreover, since auxiliary lines were drawn as part of many of the proofs in our textbook, how did one know that it was possible to draw the figures as shown? The textbook didn't say. Wasn't therefore the whole carefully built up structure of Euclidean geometry a fraudulent imposition?

This is where physical empiricism came to my rescue (but not yet to the rescue of Euclidean geometry as presented in my text). I loved to play around with a ruler and compass. I determined to recreate the construction given in the "proof." I carefully drew a scalene triangle ABC (i.e., one with unequal sides). I carefully constructed the perpendicular bisector of side AB. I carefully drew the bisector of the angle C. I sought their intersection.

Zounds! The situation was not at all as depicted in the book. In the figure in the book, the two lines intersected *inside* the initial triangle ABC. On my sheet of paper, they intersected *outside* the triangle and the subsequent con-

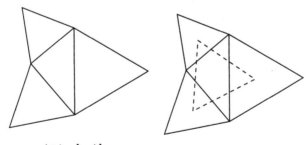

A triangle with
three equilateral triangles
constructed outwardly
on its sides.

The centers of the equilateral
triangles themselves are vertices
of an equilateral triangle.

Napoleon's Theorem.

struction could not be carried out. I drew a number of scalene triangles, and the result was always the same. The "proof" collapsed.

This discovery was enough for me at the moment. In one of my more advanced high school mathematics classes students had to make oral presentations, and I selected this paradox for my exposition. I was firm as to what my ruler and compass construction told me. And though I spoke about the relation between geometrical figures and logical deduction, that part of my talk, I would have admitted even then, was pretty fuzzy. Apparently, one must be prepared to distinguish logically between the inside of a triangle and the outside, and clarify the idea of one point lying on a straight line between two other points.

I also did not realize at the time that what is paradoxical can almost always be turned around to express a positive theorem. In this case: in a scalene triangle, the perpendicular bisector and the bisector of the opposite angle must *always* intersect *outside* the triangle. In mathematical lingo, this type of statement is called a triangular inequality (of which there are many). More than this must occur as to the relative positions of the auxiliary lines, but I'll leave it at that.

Later, poking around in other geometry texts, I found three or four putative proofs of impossibilities based on other misrepresented constructions. As for an explanation of the general relationship between Euclidean deduction and the reliance on drawn figures, I still had to learn quite a bit more.

Part II

Undergraduate Years

Little Did I Know

When I entered Harvard in late September, 1939, what mathematical knowledge did I carry with me? As I have said already, I took Algebra I, Algebra II, Plane and Solid Geometry, and Trigonometry in high school. Then there were the subjects I had taught myself: first year calculus, analytic geometry, some determinant theory and a bit of point set topology culled from E. V. Huntington's *The Continuum*. In addition, there were a few things I had worked out for myself, such as the Newton formula for polynomial interpolation.

My senior graduation present from my brother was a Keuffel and Esser loglog slide rule of which I was very proud. (Many students bound for the sciences had a slide rule, but few owned a loglog!) I still have it, though now, in relation to scientific computation, as someone said in a different context, it stands as the babbling syllables of an infant to the orations of Cicero.

This knowledge was enough to score me high in mathematics on the college board exams, and ultimately to admit me to both Harvard and MIT. I have to confess that I did poorly in the English boards, my literary knowledge and abilities then being close to the babbling level. I used to have a theory, based on some empirical evidence, that if one's brains were wired up for mathematics and science, they were wired down for writing, literature or history. But my own experience at Harvard soon made hash of a part of that theory. I think now that the distance from the bush of a provincial high school to symbolic or verbal sophistication of whatever sort is but a micron compared to the distance mankind has traveled from the primal slime to the bush.

A recent story rewarmed a certain part of my theory. I was wandering in the halls of the Technical University of Berlin, attending the 1998 International Congress of Mathematicians, when a man came up to me and introduced himself as Professor Such and Such, retired now from Thus and So university in Europe. I didn't really know the man, but I had the feeling that I had met him somewhere before, perhaps even at Thus and So. "What are you doing now that you are retired?" I asked him. His answer was, "I'm working on a study on why professional mathematicians are, on the whole, so unintellectual." I responded, "Well, good luck with your study. I hope you find out. I've had the same feeling for years, but it would have been professionally disadvantageous for me to go public with it."

At the moment of matriculation, my knowledge of mathematics, though sufficient, as I have said, to get me high college board grades, was in no way remarkable when compared to the knowledge of some freshmen who came from prep schools such as Phillips Andover or from elite public high schools such as the Bronx High School of Science. What carried me forward to a professional career in mathematics was not so much the textbook knowledge— though I shouldn't underestimate the necessity and the difficulties of absorbing such knowledge—but an ability and a desire that emerged early to pose my own mathematical questions, to dig deep, to conduct my own mathematical experimentation and ultimately to answer the questions in my own, often clumsy, fashion; in short, to do original research.

I recall getting the key to the mathematics library (special privilege!) which was then housed on the top floor of the Widener Library. Seeing shelf upon shelf of advanced math books, I wondered whether I could or would ever need to know all that these tomes contained. I selected a book at random and flipped through it. I recall it was in French, it was thin, and it dealt with something called Cremona Transformations (Luigi Cremona, 1830–1903). It might have been Godeaux's *Les transformations birationnelles de l'espace*. I put the book back gently and said to myself, "Perhaps I'd better learn some mathematical French." The years passed; I've learned some French and some Russian. I've learned a good deal more German, but very little about Cremona Transformations.

Freshman Logic

The social anthropology of the Harvard undergraduates in the years 1939 to 1943 has been described brilliantly by my former roommate Myron Kauffmann in his 1959 novel *Remember Me to God*. My personal taxonomy was as follows. There were students from public high schools and students from prepschools. There were grinds and there were the clubmen. There were athletes and there were the physically inept. (How did I ever manage to scale an eight-foot slippery wall as part of "military training" after war was declared in 1941?) There were those who lived on campus and those who commuted to classes. There were the rich and there were the scholarship students. There were the politically active and the politically indifferent. There were the left-wingers and the right-wingers. There were the drinkers and there were the non-drinkers.

J. Manville Goodheart, just across the Yard in Hollis Hall, was a bookie with an undergraduate clientele. J. Manville had a radio in his room; it was necessary for him to know the racing results. Most students didn't bother to have one. I remember listening to Churchill's famous "Blood, Toil, Tears, and Sweat" speech over J. Manville's radio as I passed by his room on the way to a friend. (May 13, 1940).

Finally, there were the bright and there were the not-so-bright; and among the bright, there were those who would say "I haven't cracked a book since midyears and I don't know how I'll keep my place in Second Group." All mixtures of these categories were possible, though some were very unlikely. I myself was a bright public high school campus resident, on scholarship, a student for whom gentlemanly C's would not have sufficed to keep the money flowing.

One of the first things I learned as a freshman was that no matter how bright you are in a particular field, or how bright you think you are, there is always someone around who is brighter. Always. (Think of the problems that have stumped Einstein. Someone "brighter" may solve them or someone "dumber" might.) Coming from a relatively small-city high school, I was in mild shock when I realized the brilliance and the accomplishments of some of my class-mates. Some of them had already written three unpublished novels. Some had made movies. Some had scaled whatever famous mountain had been first scaled in the '30s. I learned fairly rapidly to accommodate to this fact of life, to rise to my own appropriate level, and I think this accommodation has over the years spared me a good deal of competitive envy and anguish.

Brilliance is one thing, fire-in-the-belly is another, and the ability to carry through plans is yet a third; these three do not come free. There may be a price that has to be paid. I learned that from my freshman roommate Peter Simon.

Peter was older than I by three years. My, what a difference three years make when one is sixteen! He was an exotic, having grown up in the English/American colony of Shanghai. I was a provincial. He was well seasoned and a bit flashy, wore baggy tweed trousers and sported a long key chain and pointed shoes with leather heels (slightly zoot, as they said in those days). By compari-son I was a greenhorn, right off the Boston and Maine Railroad from Ballardvale Junction. Despite these differences, we got on splendidly and Peter undertook to educate me.

Part of our compatibility, I suspect, was due to the fact that, as I have already pointed out, I was a born skeptic. I've read that all children are skep-tics; how else to explain the constant questioning "Why, why, why?" My skep-ticism has been deeper and has lasted longer, and for this I put the blame on Ernst Haeckel.

Haeckel (1834–1919) was a German zoologist and early defender of the theory of evolution. At the turn of the century, he wrote a book, translated into English as *The Riddle of the Universe*, in which he put (not very satisfactorily) evolutionary theory to the purposes of philosophy. At any rate, this book was hanging around our house, and at the age of eleven I looked into it. I couldn't make sense out of it; much too technical. But one sentence I understood: Haeckel said it was not unreasonable for the ancients to have worshipped the sun considering that everything that takes place on earth is due ultimately to the beneficence of the sun.

[An aside: I did not realize until recently the strong influence that Haeckel had on educated German youth of his period. In Gerhart Hauptmann's play *Die Einsamen Menschen* (*Lonely People*, c. 1895), a young man influenced by reading Darwin, Haeckel, and Emil Du Bois Reymond—authors mentioned specifically in the dialog—is put on a collision course with his conventionally religious parents. The play ends in tragedy.]

I cannot now follow the stream of thought that led from that sentence to the strengthening of my satiric skepticism, but it was fully developed by the time that I shared Thayer Hall 13 with Peter Simon. In one of the early bull sessions in our room, the question of formal logic was bandied about.

On matriculating as a freshman, my experience with formal mathematical logic amounted to this: I knew that it is possible to draw false and ridiculous conclusions from geometrical figures that have been drawn imperfectly; I had heard of the *Principia Mathematica* by Whitehead and Russell. When I was a senior in high school, I asked for it at the Lawrence Public Library. They did not have it. I put in a request for it, but by the time it arrived I was already a college freshman and my family had moved out of Lawrence to Manhattan. As I made out the request slip, no oracle whispered in my ear that more than a half century later, I would deliver the first Whitehead Commemorative Lecture at Imperial College in London.

One more thing: I wondered, if mathematics were mere logic, how logic, a product of human thought, encapsulated in certain scratches on a piece of paper, could compel, for example, the three angle bisectors of a triangle to meet at a single point when you drew them. One system of scratches controlling another system? Put it another way and more generally: how can my abstract thought tell the universe what it should do or will do? Or have I put the cart before the horse? Perhaps by evolutionary processes, my abstract thought derives from and must agree with how the universe actually behaves.

Returning to the bull session, hardly three opinions had been offered by the group when a couplet popped into my mind (I was then taking German) :

> *Eine feste Burg ist unsere Logik*
> *Sie besteht aus reiner Hodgik Podgik*

> (A mighty fortress is our logic
> It consists of simple hodgik podgik.)

Peter was enchanted by this mildly skeptical and early anti-positivistic statement on my part. My status in his eyes went up enormously. He pronounced me the Luther of Middle Entry Thayer Hall. And it was only the middle of October.

As regards fire-in-the-belly. Peter signed up for the basic freshman course in English literature. He needn't have. He knew it all anyway. In the spring semester, a prestigious prize for the best essay was announced. Peter decided he would compete and win. He selected his topic: the poetry of Robert Burns. The next thing I knew he had taken out of the library forty books on Burns, piled them on the radiator and was beginning to digest their contents. This bravura performance boggled my mind and indicated the depth of scholarly engagement that can exist.

From March until the end of the semester, Peter immersed himself totally in Burns. Everything else was forgotten, neglected: meals, dates with girls, other courses. His essay won the prize. He flunked all other exams, and was put on a serious level of academic probation. How stupid of them, I thought; how very, very stupid.

Peter needed a summer job. His mother had a position with one of the Manhattan magazines and she came to know Bennett Cerf, Publisher of Random House books. Cerf was a wit and a collector, i.e., he collected jokes and wrote and published joke books. A standup comedian with intellectual accomplishments. Peter wrote a funny script asking Cerf for a summer job at Random House. He cut a disc (this was before tapes were around), and sent the disc to Cerf. In reply, he got a funny letter from Cerf saying No.

In the middle of the Summer of 1940, Peter died in a climbing accident and I and my classmate Ricky Leacock (cameraman for Robert Flaherty and an early advocate of Cinema Verité) were at the cemetery to see him buried. What faith, what philosophy, what logic, what equations of mathematical physics can measure the value of enthusiasm or can answer the why's of life?

Why I Didn't Take Philosophy A

As an entering freshman I wondered which of a wide variety of courses to take. Mathematics and astronomy, certainly. English? Well, I couldn't get out of that. Physics, of course. That left one more course that might be taken. Ah yes, those damn "distribution" requirements without whose fulfillment one could not graduate (like the swimming test). Art? Never. History? No way. Psychology? Yukkk. Philosophy? Well, maybe. Let's see who's giving the course. Ah. Prof. Raphael Demos.

The name Demos, which I took to be of Greek derivation, reminded me that Philosophy A would probably be all about Plato and Aristotle and those guys. So I went to the reference room of the library, pulled down an encyclopedia, and looked up Aristotle. Finding a very long article, I read at random.

There are four causes, said Aristotle, the material cause, the formal cause, the final cause, and the efficient cause. As I understood it, the material cause was simply the material involved, e.g., a piece of paper, a feather. The formal cause was the form or the shape of the material. The final cause was the purpose or the goal or the outcome of whatever was under discussion. Lastly, the efficient cause was whatever set the whole business going in the first place.

I wrestled with Aristotle's four causes and they seemed in no way to be connected with my experience; a world of ideas I then saw no reason to enter. A week before, I had just been driven down to Cambridge with my bags by my older cousin Shirley. Her father, my mother's brother, was a man whom I liked very much. It was said by all those who knew him that he would never go anywhere outside of Lawrence, Massachusetts and never had done so for years and years. One day I asked him why this was the case, and without any hesitation he answered, "I need to empty the drip pan under the ice box every day."

This seemed sufficient cause for his action, but I could not fit this explanation into Aristotle's scheme. Was the material cause the ice box, the cake of ice, or the drip pan? Was the formal cause the cubical shape of the first two or the cylindrical shape of the latter? Was the final cause that my uncle never left Lawrence? Was the efficient cause the fact that my aunt was not in condition to empty the pan or that the flat they lived in didn't have a drain hole?

I decided that Philosophy A was not for me. My damned distribution requirements were not fulfilled until my senior year (with advanced German), and the residue was a deep and abiding suspicion of all embracing systems of thought. Such general philosophy as I now know, I picked up "behind the garage," so to speak.

The Skies Were Limited

I did not enter Harvard with the idea of majoring in mathematics. I knew I wanted to be an astronomer. The night skies in the Tower Hill area of Lawrence had a sufficiently wide horizon and were sufficiently smogless to observe the stars. I took to gazing at the stars first as a hobby and then a bit more seriously. I got myself a star book and got to know and to identify most of the visible constellations. This took the better part of a year, because on any particular day, the whole of the potentially visible stars at latitude 42 are not out between sundown and, say, midnight when it was time to tuck in. I got to know the major stars in the individual constellations, Betelgeuse in Orion, Vega in Lyra, Antares in Scorpio, Ursa Major and Minor of course, Algol in Perseus, Altair in Aquila, their colors and their brightnessess. And numerous others. If part of the heavens were clouded over and there was a hole in the clouds, I could generally tell what constellation was peeping through.

I could identify the planets and knew their paths through the zodiac. More mathematically, I knew the distinction and the computational relations between civil time and sidereal time, between the calendar month, the synodic month, and the sidereal month. I owned a cheap pair of opera glasses, and found the Pleiades and the great nebulas in Orion and Andromeda. I dreamed of owning a two-inch refracting telescope, which Meyrowitz's optical shop in New York City sold for $90. But $90 was almost one-quarter of a year's college tuition, and the dream of possession remained a dream.

It naturally followed that as a freshman I would sign up for the introductory course in Astronomy. The course was taught by a wiry Dutchman by the name of Bart J. Bok, thirty-three years old at the time, an assistant professor and a man who subsequently made a great name for himself in Milky Way studies. The 90-inch telescope on Kitt Peak, Arizona, is named after him.

There were two lab assistants in Astronomy 1 whose name I would probably recall under hypnosis; the older man was an expert in the computation of comet orbits. The doyen of American astronomers, Harlow Shapley, had an office just down the hall from the classroom. A short, roundish man, he would pop in and out every now and then from a huge circular desk on which he worked on topics in rotation.

We used the old 15-inch refractor on Observatory Hill, Cambridge, and I took a picture of the moon. We traveled 20 miles to the west to Harvard, Massachusetts, where the university's great 60-inch refractor was located, and I saw how "real" astronomers worked: wearing long underwear, overcoats and ear muffs in the freezing temperatures under the dome. Observatories cannot be heated; the heat waves would result in distorted images.

The freshman course soon left the elementary area of descriptive astronomy that I had taught myself and the positional astronomy (celestial mechanics) that I was willing to learn. We went on to astrophysics. There I found myself in shock. I was used to deriving and proving things from axioms in geometry and in calculus, and I could not figure out the logic of such things as the Russell-Hertzsprung diagram or the mass-luminosity relationship for "main sequence" stars. Here they are, the book and the Bok seemed to say. No proofs in the mathematical sense. No derivations from the fundamental laws of physics. (If there had been, I probably would not have understood them anyway.) Just observations, plottings on graphs paper; believe them, learn them, use them, make deductions from them. If Bart Bok had been Ring Lardner and I had had the guts to raise the issue, "Shut up," he would have explained.

I began to see that I did not have the mindset of an astronomer. Mathematics, for all its rigidities, for all the doubts I had as to its logical foundations, or logic itself, mathematics was then where I was at. I settled for a B in the course and left astronomy forever. Professor Bok went his way and I went mine back to mathematics. Today black holes interest me only insofar as they are a mathematical concept. The rest of their romance, I'll leave to the imagination of the sci-fi writers.

Undergraduate Mathematical Life:
Harvard 1939–1943

In my sophomore year I had second year calculus with Marshall Stone and projective geometry with Oscar Zariski. Junior Year: advanced calculus with George David Birkhoff, probability and statistics with E. V. Huntington, abstract algebra with Joseph L. Walsh, mathematical logic with Willard Van Orman Quine. Senior year: complex analysis with David V. Widder and philosophy of science with Philipp Frank. I was entitled to take "course reduction" and availed myself of the option. Less is more, I've always felt.

With the exception of my first year calculus teacher, all my undergraduate mathematics teachers were full professors. All the names I've just mentioned were stars in the mathematical heavens but I came to realize that only gradually. Naturally some stars burn brighter than others.

The lecture system was in full force. Depending on the individual, the professor lectured from notes or ad-libbed from "the top of his head." The students took notes. In either case, when the professors went to the blackboard, all of them managed to give me the impression that they were creating mathematics *ab ovo* right there in front of me. I came to realize that was a part of the theatricality of higher mathematics education. After World War II, when less sartorial formality was demanded in the classroom, the theatricality intensified. The younger professors would take their coats and ties off, light up a cigarette and pretend that the theorems and the proofs they had committed to memory and then imparted as *de novo* material required working clothes and not the wing collars that persisted into the mid-Thirties.

My sophomore year tutor was George Mackey, with whom I read group theory. My junior year tutor was D.V. Widder, with whom I read Knopp's *Infinite Series* and Hardy's *Pure Mathematics*; my senior year tutor was Lynn Loomis, with whom I wrote a senior thesis based on the work of the great French mathematician Joseph Liouville.

From time to time I went to the Mathematical Colloquium (this was for the pros) and to the undergraduate Math Club. As a junior, I corrected calculus papers for Marshall Stone's class and for Instructor Al Whiteman's class. As an undergraduate corrector, I was paid perhaps a dollar per student per semester, and it was well-appreciated pocket money. To make additional pocket money, I tutored students who were failing in their math courses: usually clubmen. I also did *pro bono* math tutoring of kids from "across the tracks."

I can recall the exact moment when I began to feel grown up—almost professional—mathematically speaking. From time to time, instead of bringing Whiteman's homework papers to his classroom, I would deliver them to his house. Three members of the department, Al Whiteman, Irving Kaplansky, and Leon Alaoglu, all bachelors then and all prestigious B.O. Peirce Instructors, shared an apartment off Kirkland Street. One day, I found Kaplansky reading *The Mathematical Reviews*. This was (and is) a periodical that gives short summaries and occasionally critical opinions of the latest research papers in the field. The realization that there was such a publication blew my mind. Mathematics ceased forever to be a short, sweet and completed set of definitions, propositions, formulas, and procedures that could be summed up in two or three textbooks. It became a living organism of developing ideas.

What Path Shall I Take in Life?

This story begins with English A in 1939. Mandatory for me. I couldn't get out of it. My writing was terrible. As I've already noted, I used to think—perhaps it's still true by and large—that there is an inverse correlation between those who can write well and those who can do mathematics.

English A consisted of reading and discussing the pieces in *Five Kinds of Writing*, an anthology comprised of (you guessed it!) five kinds of writing. In addition, we had to turn in a thousand words a week. The course grade was based essentially on these words. These thousand words were murder to me. I couldn't think what to churn out. If the assignment was to criticize Henry James' *The Turn of The Screw*, I lacked the sophisticated vocabulary, the current chic point of view. If I were to say simply that I didn't get it, that I couldn't follow it (which is still the case), I could express that honest opinion in far fewer than one thousand words. I cheated on the number of words I used to turn in, and the little that I was able to squeeze out I turned in late.

January came around, (this was a one year course), and I received a D for the semester. A serious conference with Mr. Murphy, my English A Section Man (postdoc, in modern terminology). "Davis, you better shape up... or else..."

The Spring Semester, 1940 came, and with it came my salvation and my shaping up. They emerged from the interior of the Old Howard, a burlesque house in Scollay Square, Boston. All undergraduates went to the Old Howard as part of their socio-anthropologic teenage initiation rites.

Before the show started, the "candy butchers" came out and sold candy and risqué pictures at enhanced prices. The butchers had a certain spiel and a singsong that employed super-refined language that they would repeat over and over again. (The closest imitation today to this sort of super-refinement is the language of the airline flight attendants: "Do have a pleasurable day in Boston, or wherever your final destination may take you.")

My ear caught the spiel and the singsong. I wrote it down in a thousand words, and turned it in on time. Result: an A. From that point on there was no stopping me. I went from strength to strength, wrote political satires, spoofs of undergraduates and professors, and got an A for the semester and B for the whole course. Thus, if it weren't for the strippers Hinda Wassau, Ann Corio, the redheaded bombshell Margie Hart, advertised as "the Poor Man's Garbo," and all their sister ecdysiasts, I might have flunked out of Harvard. Out of the Fall comes the Redemption: a Chassidic doctrine.

A half century later, I read a very substantial biography, just published, of Jack Yeats (William Butler Yeats' father), written by one William Murphy. From the bio on the dust jacket I concluded this William Murphy had been my section man. I wrote him and said "Thank you, Mr. Murphy. Years ago you saved my life." He acknowledged my letter very kindly, writing "I'm glad I did."

To return to the fork in my life's path. Having been raised from the dung heap of incompetent prose to sit at the feet of the literary kings of the world, I took the next higher writing course in my sophomore year. I got along splendidly with my section man, Albert Guerard, Jr. I wrote short stories—now lost, thank goodness. I admired Sherwood Anderson, who was then an acceptable author to admire but, still, did not carry the cachet of Ernest Hemingway.

I became a short story buff, collecting old copies of *Story Magazine* from the Third Avenue used book stalls. I read *Best Short Stories of 193..*, edited by Edward O'Brien. By the end of my Sophomore year, I had two loves: mathematics and writing. For several months I felt like Buridan's ass who faced a fork in the road, both branches of which seemed equally attractive. But something happened in the course of taking English A1 that forced a decision.

While I was doing well with my writing, very well, in fact, I was thrown in with a few students for whom writing really was their whole life. Was writing really my whole life? Was I ready to be an impoverished artist for the sake of a few well-turned sentences? Did I think that I had unique moral observations and Voltaire-like mockeries that would rescue civilization from its follies? Did I have the literary fire-in-the-belly of fellow undergraduates such as Howard Nemerov or Norman Mailer who would later rise to great heights of literary prominence? Or even of my ill-fated roommate Peter Simon? I thought not.

What gave my literary aspirations the *coup de grace* was this: by June 1941, it was clear that the USA was going to enter World War II on the side of the United Kingdom (First "Lend-Lease," then "Bundles for Britain," then en-

try). Undergraduates who majored in science could probably get a draft defer-
ment; those who majored in the humanities probably could not. Even as I took
advantage of my math major, the brutal conjunction hit me: that mathematics
was one of the many handmaidens of war.

Elève de Frank T. Hubbard

It happens not infrequently: what begins as a firm and devoted friendship tapers off and becomes attenuated as time goes on and as circumstances change. The world is large; one must go where the opportunities are , and even in these days of superb—and almost three-dimensional—communication, geographical separation is not easily bridged. If one is lucky, one can learn "what happened then," and update but incompletely the course of the lives involved. In the case of my friendship with Frank Hubbard (1920–1976), I have been lucky—due, in part, to the Web.

My first encounter with Frank was not propitious. In the fall of 1940, I was eating lunch in the Dunster House dining room, sitting by myself at a table for two by the wall. A short, slight fellow with sandy colored hair came by and asked whether he could join me.

I nodded, and he sat down. I found him gruff, a bit ill-mannered, a bit cranky, impatient. The waitress brought him his lunch. He gobbled it down, and whisked off before I had finished my dessert. Other than observing his manner, all I found out on that occasion was that he was a junior (I was a sophomore.) I don't really like that guy, I said to myself after he'd gone.

The same lunchtime encounter occurred later. The same gruffness, but more conversation this time. And then in the weeks following, lunch again and again. I began to realize that underneath the gruffness, there lay hidden a very interesting and generous personality. A sweet melon with a tough skin? No, I doubt if anyone would have called Frank sweet. A rough diamond, maybe. Our friendship grew, at first casually, and later quite substantially.

Frank was majoring in English. His dormitory room was on the first floor of G-Entry with a huge window that overlooked the enclosed quadrangle, which itself was behind the formal gates of Dunster House along Memorial Drive and the Charles River. I was in his room frequently. What a mess! Books and papers scattered everywhere. More than that, bicycles and parts of bicycles under repair, all over his rooms. One had to negotiate one's path carefully to find some place to sit down and talk to Frank while he was puttering around with wrenches and screwdrivers and greasy hands.

In the soft spring weather with his windows open, people would talk to him from outside. He played the violin. He practiced regularly, and returning from supper across the quadrangle to E-Entry where I lived, I could see him and hear him sawing away on his fiddle. Bach or Vivaldi; something like that. He once told me he had no interest whatever in classical music after Beethoven, and in a later conversation, he drew the line at Mozart.

I tabbed him for a future professor of English literature; he seemed to be headed in that direction. But there were three passionate interests residing in his breast: music, scholarship, and craftsmanship. How he was able to combine all three successfully, I would find out over the coming years. Intelligent, forceful, scholarly, a craftsman with the highest standards, enlightened; and living musically in the seventeenth and eighteenth centuries, he was the closest person to our third President, Thomas Jefferson, I've ever known.

Perhaps a year afterwards, Frank was courting (to use a term that was archaic already in the 1940s). His girl was Ruth Hoffman, a physician's daughter, driven out of Vienna, and a freshman at Radcliffe. I recall a meal at Ruth's family home, somewhere in the Boston suburbs, her grandmother serving us a rich Viennese broth.

World War II separated us. I believe he served in a meteorological unit, and at one point was stationed in Baghdad. We were both married and I was then at the laboratories of the National Advisory Committee for Aeronautics (NACA) at Langley Field, Virginia. During this wartime separation we did not correspond.

Peace restored, but not yet the ravaged cities of Europe, we were back together again in the fall of 1946. He was a graduate student in English Literature and received a MA in 1947. I was a Ph.D. candidate in mathematics. He and Ruth lived in an absolutely ramshackle house not far from Central Square in Cambridge. Some of the rooms resembled his college dorm room, strewn with parts of instruments, books, woodworking equipment, parts of bicycles. Frank was making a violin from scratch, so in addition, there were clamps and planes and saws and gluepots, not to mention specialized jacks and jigs for violin making. Several violin backs in the process of construction were strung up.

Our friendship renewed. I was, in those days, and beginning in high school, a fairly enthusiastic recorder (blockflute) player. I owned a soprano and a bass recorder, the latter devolving to me from Bimbo Wedgwood of the English Wedgwood family, his first name now escaping me. Frank soon organized a recorder quartet. I generally played the bass, and he the soprano.

The quartet membership floated a bit. At one point, it included I.I. (Izzy) Hirschmann, who later had a distinguished mathematical career at Washington University in St. Louis, and Alison Lurie, who became a professor of English at Cornell and a very successful novelist. When Izzy was around, he played soprano, and Frank would then improvise or double on his violin, creating a quintet. We met regularly in Frank and Ruth's living room, and we always played pieces from the baroque recorder literature.

The high point of the quartet came, if I recall correctly, in December of 1946. Frank was taking a course in Anglo-Saxon. He had formed a friendship with Professor F. O. Matthiessen (1902–1950), a tremendously popular member of the English department. Matthiessen asked him whether he could arrange a Christmas concert for his class, baroque recorder music being appropriate to the subject matter of the course he was then giving.

Frank agreed. The quartet was enthusiastic and we worked up a program. On an evening before the Christmas vacation, our quartet performed with elan and eclat in one of the large rooms of the Harvard Student Union (the performance, luckily not taped, though tapes were already around). When we players had finished, Professor Matthiessen invited us all to "partake of the wassail bowl." I remember distinctly that he pronounced this Middle English word as "wa-sale" ("wa" as in water), and I have ever since assumed this to be the correct way. The quartet did not survive its peak performance; time, place and other obligations pulled us apart.

Frank abandoned his Ph.D. plans (if he ever had them). In 1947 he and Ruth traveled to cold, rationed, half-destroyed London, where he became an apprenticed harpsichord maker at the famous Arnold Dolmetsch Company in Haselmere, Surrey, England. In 1948 he was apprenticed to Hugh Gough in London as a harpsichord maker. By 1949, he was back in Boston, and together with Bill Dowd, a fellow Harvard student and another harpsichord enthusiast, he founded Hubbard & Dowd, Inc. with a workshop in a loft in downtown Boston. (Hubbard Harpsichords, Sudbury, Massachusetts, recently celebrated its fiftieth anniversary.)

In 1951, with my degree in my pocket already for some months, I was back in Cambridge after a very brief hitch at the U. S. Naval Underwater Sound

Frank Hubbard leaning on a 1750 english harpsichord by Jacob Kirckman.

Laboratory in New London, Connecticut (Now disbanded). The NUSL is where I first heard that plans were brewing for a satellite, and I was incredulous. In Cambridge, I was teaching freshman calculus at MIT and working also in one of MIT's large government-sponsored defense development projects left over from World War II.

Hubbard and Dowd Inc. prospered. They hired young men with cabinet-making training. Their fame grew. The interest in harpsichords grew, often in people who wanted the pleasure and the pain of making one themselves. H & D became one of the leading harpsichord makers in the USA. (Harpsichords were not produced by the standard commercial manufacturers of pianos such as Steinway or Baldwin.)

I visited their shop often, just to talk with and watch Frank at work: designing, measuring, consulting documents, cutting, stringing, arranging for gold leafing. After an hour or so of my hanging around, he might brusquely pick up his violin case from the corner, and run off, saying that he'd promised to play in such and such a group. At one point, while he was laying out the innards of a harpsichord, he said he had run into a certain geometrical problem there and could I solve it. It was simple enough; a bit of trigonometry et voila!

I finally thought to myself, enough of just hanging around. Perhaps I can make something for myself in Frank's shop. The three years of woodworking I'd had in grammar school left me with a hankering to do something with tools. I broached the matter to Frank, and he responded generously. He would guide me, advise me, teach me the use of the various electric shop tools. He taught me the use of jigs—auxiliary tools for tools. But I would have to do all the woodworking myself.

I drew up a plan for a low and sturdy coffee table—a harpsichord would have been totally beyond my capabilities, patience, or need. Five pieces all in all, to be screwed and glued but not mortised or tenoned. Frank approved my plan. From his lumber supply, he selected a large, rough piece of Honduras mahogany and sold it to me for ten dollars. And then I was on my own.

I put in the odd hour now and then, and it took me a number of months to complete what, structurally, was a very simple job. Frank would come around every once in a while and take a look. "Needs more sanding there. You've got to put a curve on the edges. Here. Use this tool." No careless, slipshod work out of his shop.

The day came, finally, when I thought the job was done. I oiled the surface and rubbed it down. I offered my work for his approval. He looked it over, right side up and upside down and approved. And as a tacit sign of his approval, he got out a special die and, turning over the table, tapped the die into the underside of the table where it cut the words "Hubbard and Dowd" into the wood. The table is still as sturdy as ever, and the words are still visible.

In 1952, I left for Washington, D.C., and this is when our getting-togethers but not our friendship tapered off. The words that follow now are derived partly from direct knowledge and partly from the website of the Hubbard Harpsichord Co.

Frank's marriage to Ruth broke up. There were no children. His partnership with Bill Dowd broke up and each went their individual ways making harpsichords, clavichords, etc. He took on apprentices and filled them with tremendous enthusiasm. His apprentices went their own ways and set up their own shops. He wrote a book describing classical instruments: *Three Centuries of Harpsichord Making,* Harvard University Press, 1964. He lectured. He repaired old instruments in museums. He put out and sold successfully a line of harpsichord kits. (Needs some assembly!)

After I had taken a job at Brown in Providence, much closer to Boston than Washington, there were a few more get-togethers. He had by that time moved his shop from Boston to Waltham, where he occupied the stables of the historic Lyman Estate. My family and I visited him several times there, and on the last time, I observed that one of the stable rooms was filled with deconstructed organ pipes from an instrument he was restoring.

Frank was a good friend, a man I admired tremendously, but never idolized. I never was able to get beyond the image of a young man playing wildly on his fiddle before an open dormitory window of Dunster House. I was not aware of how much idolization of Frank went on in the larger world until one day in the late Seventies. My wife and I were driving back from a vacation at Penobscot Bay in Maine, and we decided to break our return trip by stopping off in Portsmouth, New Hampshire, and taking a look at Strawberry Banke, a colonial village reconstructed after the manner of Williamsburg or Sturbridge. In the "craft" part of the village, we saw a shop with a shingle that read "Instrument Maker." Coming closer, we read the words " So-and-so, Elève de Frank Hubbard."

I, too, was an élève de Frank Hubbard.

How I Was Turned off Formalism

The story goes that Dr. Samuel Johnson once refuted Bishop George Berkeley's denial of the material world by kicking a stone with his foot and saying "I refute it thus." My story is a bit more complex.

Mathematical formalism maintains that mathematics consists simply of a sequence of symbols that are to be processed in accordance with certain rules of manipulation. The symbols and the manipulation have no meaning or existence except insofar as they are able to be connected with other symbols or occasionally be made to connect with processes in the non-mathematical world.

Most practicing mathematicians care very little about discussing the philosophy of their subject, but they work unconsciously with a philosophy of Platonism. Platonism, briefly, asserts that all mathematical ideas and constructs are independent of people. Pi (= 3.14159...) is in the sky. If the shortcomings of Platonism are pointed out, mathematicians usually fall back on formalism.

How was I turned off formalism? I like to tell people it was due to Charles Lindbergh (1902–1974), the famous Lindy who in 1927 flew solo across the Atlantic in the *Spirit of St. Louis*. But patience, there's no point in spoiling a good story by this sort of reductionism.

In the fall of 1941, in my junior year at Harvard, I took a course in mathematical logic (Mathematics 19) with W. V. O. Quine (b. 1908), then associate professor of philosophy, and later to become perhaps the principal and most honored American philosopher of his generation. There were about ten students in the course, whose papers were corrected by a certain Mr. Berry who was Quine's assistant.

I remember Quine (he was only thirty-three then) as a crisp, rather humor-less, no-nonsense lecturer with a hairbrush moustache. A midwesterner, he was educated at Oberlin and Harvard. Years later, I had lunch with him at the American Academy of Arts and Sciences in Cambridge, and found him to be a very tough man to amuse. I like to say of myself that while I'm serious, I'm not somber. Quine was both.

Quine had studied the mathematical philosophy of Alfred North White-head and Bertrand Russell, and been exposed to the philosophers of the Vienna Circle as well. He had travelled to Europe and come into contact with its prominent members. Quine had just published a text, *Mathematical Logic*, which was kind of a short and updated version of Whitehead and Russell's famous *Principia Mathematica* and embodying a number of his own researches. This was the text we used for Math 19: Mathematical Logic.

The book is full of sentences written in the special and not altogether standardized notation of logic such as:

*144. $\vdash \ulcorner (\alpha)(\phi \lor \psi) \supset . (\exists \alpha)\phi \lor (\alpha)\psi \urcorner.$
  Proof. *102 (& D4, 8) $\vdash \ulcorner (\alpha)(\sim \phi \supset \psi) \supset . (\exists \alpha)\phi \lor (\alpha)\psi \urcorner.$ (1)
    *100, *123 $\vdash \ulcorner [1 \equiv]144 \urcorner.$

When we came to study page 157 of the text, Quine threw back at us students a logical bombshell that had been thrown at him. The axiom of membership listed as *200, taken with the other axioms in his book, were discovered by J. Barkley Rosser (professor of mathematics at the University of Wisconsin in Madison) to lead to a contradiction. Wow!

Now contradiction is considered by mathematicians to be THE PRIMAL SIN, and therefore a certain fraction of the subsequent pages of Quine's book stood there in naked error. A hasty fig-leaf job had to be done, and the class spent the next few weeks penciling in which of the subsequent statements were valid, which were invalid, and how the invalid statements could be legiti-mized by the outright adoption of seven later theorems whose given proofs involved *200.

Well, in committing a logical booboo, Quine was following in the steps of his great idol Bertrand Russell, who had to patch up a paradox of the theory of sets with a rather unsatisfactory "theory of types" that had problems of its own. These "bombshells" seemed to me to be great fun. "How the Mighty Have Fallen." The bombshells pointed out to me that (1) mathematics books and deductions are full of errors: they are written by humans who are error-prone, and (2) it is not easy either to get rid of errors in general or to patch up specific ones.

This theme was taken up in a grand way in the 1960s by the Hungarian mathematician Imre Lakatos (*Proofs and Refutations*), and has been parlayed into a philosophy of mathematics by Lakatos and a number of his admirers, among whom I include myself. Nonetheless, though I perceived I had been driven out of the Logical Garden of Eden, I enjoyed Mathematics 19 tremendously and I advertised its virtues widely.

More than this: In the Fall of 1942, I came to write my senior honors thesis, which was my interpretation and condensation of certain problems in finite integration. I used G.H. Hardy's books *The Integration of Functions of a Single Variable* and *Orders of Infinity*, as well as going back to the original work of Paul du Bois Reymond (1831–1889) in German, and Joseph Liouville (1809–1882) in French.

At that time, I was quite madly in love with the logical notation in Whitehead and Russell and followed, more or less, by Quine. It was a strange and perhaps mathematically chic thing to have done, but I expressed a goodly fraction of my thesis in the notation of the *Principia*. Even as I was doing it, I realized that this notation was not eliciting any new substance from the basic questions I tackled. But I did it anyway. It was a useful exercise and led me to the conclusion that the relationship between form and substance in mathematics is an exceedingly complicated matter. I don't think that anyone has yet written deeply on this dialectical split. The semiotics of mathematics is, strangely, a poorly developed subject, and I hardly understand the little of it that's around.

It was also in the fall of 1941 that I met Hadassah Nita Finkelstein on a picket line. The war in Europe: the Allies versus the Axis powers, had turned very grave for the Allies. It was clear to most of us that the United States would shortly join the Allies. But the step was not favored by all Americans. A group known as America Firsters opposed the entry of the US. They called a rally in Mechanics Hall in Boston. Charles Lindbergh, a man of great reputation who appeared to be soft on the Nazis, was their main speaker and attraction.

Groups of students from the Boston area colleges were formed to picket the America Firsters. H.N.F. was part of the delegation from Radcliffe College (In those days, Harvard and Radcliffe were quite separate). To create a brief time warp: I met her on Huntington Avenue, we talked, we dated, we became friends, we loved one another deeply, we were married, and we still are.

In the fall of 1942, we were both seniors. Hadassah, who majored in psychology, was shopping around for a third or a fourth course to take. I recommended Math 19, saying it was great and she'd enjoy it. She objected that all she knew about math was second-year calculus. I countered that for Math 19,

Hadassah Finkelstein and Phil Davis, Summer 1943.

one didn't need to know a thing; one had just to learn a few sequences of symbols, use a bit of common sense here and there, and that would get her through.

Quine gave her an A for the course. She told me later that she hadn't a clue as to what really was going on. But it didn't matter. She was able to remember a few of the logical formulas and to reproduce them on the final. Epistemology? Forget it.

Quine had given me a B for the course. And I had thought I knew what was going on. And that, dear readers, is why ever since I have disliked the philosophy of formalism.

Whitehead and Russell

I matriculated in September 1939, and Alfred North Whitehead (1861–1948) had retired from his third career (Cambridge University; Imperial College, London; Harvard) a year or so previously. In his final career, he was deep into philosophy. I was not; so I would not, even had he been teaching, have taken his courses. But I saw him occasionally walking across Harvard Yard, a man then in his late seventies, stooped, carrying a stick or an umbrella, his clothes, his gait and general manner, a wonderful inheritance from the Victorian Age transplanted to a New England green. There was no mistaking this man for an American.

I had poked into the *Principia* for Quine's course; I knew Whitehead's reputation in the mathematical world; whenever I passed a certain apartment house in Cambridge, I would say to myself: this is where Whitehead lives. I learned somewhat later that in just those years, my future father-in-law had visited Whitehead on a number of occasions and had come away tremendously impressed.

Whitehead was the son of a Church of England clergyman. In his late twenties he experienced a crisis of faith. At this time also, Whitehead was shaken by the pre-Einsteinian rumbles from the physics community that the laws of Newton had only a provisional quality. He thought of going over to Roman Catholicism. He went for an "audience" with the famous John Cardinal Newman, who himself as a young man had been a convert from Anglicanism. At the same time, he met Evelyn Wade, whom he later married. The story goes that his marriage "saved" him from conversion. At any rate, as the years passed, Whitehead became the sort of agnostic who yearns for but never achieves some kind of religious outlet.

Alfred North Whitehead (Photo courtesy of Harvard University).

Bertrand Russell (1872–1970) had been Whitehead's student at Cambridge. Apart from the fact that they collaborated for a dozen years, there could hardly have been a greater contrast between two university men. Whitehead was modest, retiring. Russell was a flamboyant aristocrat (on the death of his older brother, he became Earl Russell) and a consummate snob. Whitehead was middle class. Russell was a fierce atheist; a pacifist who went to jail in 1916 for his pacifism; an advocate of sexual freedom. Whitehead was a bit bottled up, outwardly glacial but warm on acquaintance. Whitehead deplored World War I, but allowed it was necessary. His beloved son Eric was killed in it.

Russell was a great logician; Whitehead was more of a technical mathematician. Russell was a sharp (and often nasty) conversationalist, an egotist, a great wit, and a wonderful writer. I was tremendously attracted to the latter two of those characteristics and I tried to learn by imitation. Playing the sedulous ape works if practiced briefly and only once during a lifetime.

The production of *Principia* exhausted Russell's mathematical interests, so that by the time of World War I, he had abandoned mathematics in favor of his many other intellectual and social pursuits. Whereas Whitehead was hardly noticed by the outer world, Russell, in the years in which I was interested in such matters, seemed always to be in the newspapers: unconventional (in their day) views on sex and marriage, winning the Nobel Prize for literature, irritating governments, putting President Lyndon B. Johnson "on trial" for the Vietnam War, Russell was a one-man dynamo.

In the collapse of religious belief in the eighteenth and nineteenth centuries, the acceptance of Enlightenment values, and in the conflict of religion and Darwinism, the desire for firm, incontrovertible knowledge grew stronger. The desire was transferred first from religion to science, and then when "scientific laws" showed signs of instability, from science to mathematics. Cardinal Newman (before his conversion) stated that the truths of mathematics were firmer than those of dogmatic theology.

Thus mathematics in the hands of some of its creators became excessively finicky, picky. In its desire to shore up or provide firm foundations for its statements, it went on a "delta, epsilon" kick and a logic kick. (I've used in-group language here.) Delta and epsilon are standard Greek letters symbolizing tiny numbers that appeared in the course of rigorizations. Epsilon could be selected as small as one wanted. "My kingdom for a horse?" In the minds of mathematicians, the kingdom might be lost for want of an epsilon. Proof positive was the goal of much mathematical exposition, and formalization and foundationism became the sole activity of the philosophers of mathematics. The young Russell, who was an out-and-out atheist, was absolutely bitten by the bug of the promise of "secure knowledge in mathematics."

The quest for secure knowledge in mathematics was taken up later by David Hilbert, a German mathematician of enormous reputation. "We must know and we shall know," said Hilbert in defiance of a statement by Emil du Bois Reymond, a biologist of a previous generation, that there are things we can never know.

And then in the Thirties, the work of Kurt Gödel on the incompleteness of mathematical theories blasted to bits the dream of Whitehead, Russell, Hilbert, and many others. This last event occurred while I was still in elementary school. But the dream of Whitehead and Russell, the dream of absolute knowledge was part of an exclusivist view of mathematics; and it was a phase or an illness I had to go through before my inborn skepticism surfaced and allowed me to develop my own antibodies.

Independently of Gödel, I and numerous others who formed a stout minority, came to find reasons for disbelieving platonistic foundationism and challenged its hegemony.

It came therefore as both a surprise and a personal gratification when, in the winter of 1998, I was invited by Imperial College, London, to give the first

Whitehead Memorial Lecture, on the occasion of the fiftieth anniversary of Whitehead's death. What on Earth could I say that would meet with the approval of his Ghost? I decided that his Ghost, from the heights of Elysium, probably kept up with mathematical developments, and might even nod with approval when I lectured on "Mathematical Evidence" and gave my talk a social constructivist spin.

Pulled by Pappus beyond Zariski

How is new mathematics created? In many different ways, but one way is by "fooling around." Fooling around with numbers or formulas, transforming them, connecting them with other formulas, fooling around with drawings, and in these days, fooling around on the computer with mathematics packages, and most importantly, relying on your own brains to provide something new. When you fool around, then with reasonable probability, you will hit on something that you didn't know and perhaps no one else knows.

I started fooling around early. And when I took geometry in high school, the process intensified. I fooled around with ruler and compass, making drawings of all sorts. Sometimes I just used a ruler by itself. And I discovered numerous things; call them mathematical theorems if you want, but they might just as correctly be called physical phenomena.

Here is a phenomenon that intrigued me and still does. Draw any two straight lines on a piece of paper. On each line, mark three points at random. Call them A, B, C on the first line and D, E, F on the second. Cross connect these points with straight lines in three groups as follows: AE, BD ; BF, CE; AF, CD. Designate the points of intersection of the two lines in each group by P, R, Q. Then, miracle of miracles: the three points P, Q, and R will always lie on a straight line! (The line is often called the "Pappus line.")

The visual simplicity of the configuration is remarkable.

What were the aspects of this construction that I found so intriguing and remarkable? In the first place, if one generates three points "at random," so to speak, then there is no reason at all to expect that they will line up on one and the same straight line. It's true that the points P, Q, R were not generated wholly at random, one has to start with two straight lines with six points confined to

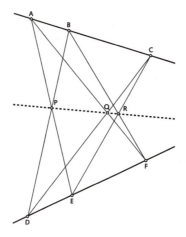

An example of the "Pappus Line."

them three by three. But still, there is a prominent random aspect to their generation. So this is a simple mathematical example of order being created out of chaos. Or at least out of semi-chaos.

In the second place, the construction requires only a ruler, not a compass. And you don't really need a ruler with inch or centimeter markings on it to draw the figure. There are no measurements of length involved either in the construction or in the conclusion. Just points and lines, their intersections and their joins. If you want, you can draw the picture using the edge of a shirt board. In technical language, this is not really a metric phenomenon.

In the third place, once the six points have been laid down, you can assign the letters A,B,C and D,E,F to them in any order you please, generating in this way different Pappus lines.

In the fourth place, the picture is quite simple. Not too many elements, not too much visual muddle. Yet, despite its visual simplicity, there are no easily discernable symmetries or self-similarities, and the reasons why P, Q, R must lie on one and the same line is not, to me at least, visually obvious.

These aspects added up in my mind to a situation of high mathematical aesthetic quality, a term that, as in art or in poetry, is very hard to define.

There was nothing like this in my high school geometry book. Most of the problems we had to demonstrate involved measurements of some sort. ("Let $AB = CD$, and $AC = BD$;" that sort of thing.) Using what I knew of geometry, I tried to prove what I had discovered and I failed. Failure can be a spur or a deterrent. In this case, it proved to be a spur, and for several years it remained on the back burners of my mind.

During the interim, in the summer of 1939, I got a job (my first) at the New York World's Fair. What an excitement for me to have come in from the sticks of Massachusetts to live and work in New York, which was then one of the great attractors for hayseeds. And the Fair? A quarter of a million visitors each day from all over the world. The Trylon and Perisphere: mathematical art and architecture dominating. Billy Rose's Aquacade. The General Motors building, where I saw the microwave oven three decades before its commercial popularity. The digital computer? Not yet born; gestating in the minds of mathematicians, logicians, and electrical engineers.

At the Fair, I worked for the Coca-Cola Company of New York selling Coca-Cola in one of their numerous booths. I wore a green uniform and for thirty-five hours a week, I got paid $17.50 less social security, which I considered a reasonable if not a princely sum. The Fair was located in Queens (some of the constructions are still standing), and I lived with my married brother and sister-in-law in their Manhattan apartment at Seventh Avenue and Twentieth Street. The subway ride on the IRT or the Eighth Avenue took me a half-hour. On my days off, and sometimes in the evening after work, I would go into the Twenty-Third Street Branch of the New York Public Library—an Andrew Carnegie donation—and read mathematics. Matrix theory and set theory were my favorites then, and matrix theory still appeals to me.

Into college and over the next year or so, I read several things about the geometrical theorem I had discovered. It was well known in mathematics. The theorem had probably been discovered over and over again by all kinds of people, but it first appears in the written mathematical literature in the fourth century *Mathematical Collections* of Pappus of Alexandria. And Pappus was able to prove it. Secondly, the theorem is now considered part of so-called projective geometry and occupies a rather distinguished position in the axiomatic development of that subject.

Before I proceed with my story, just a word about Pappus. His *Collections* (c. 320) in eight volumes is a systematic summary of most of the known Greek mathematics of the day with the addition of a few interesting things that he himself discovered. Pappus was the last mathematician we know about who was associated with and probably supported by the famous Museum at Alexandria. He was not one of your great mathematicians of antiquity, but, again, he was the last mathematician to have written anything significant before Western mathematics went to sleep for a thousand years. My high school discovery can be found as Proposition 137 of Book VII of the *Collection*. (Specialists should see Alexander Jones' wonderful Greek-English version of Pappus's Book VII with commentary.)

My sophomore year, I decided to take second-year calculus, a pretty standard choice (with Marshall Stone) and, for the sake of my interest in Pappus, Math 3: projective geometry (a whole year's course and by no means a standard one). Math 3 would be taught by a certain Oscar Zariski (1899–1986) on leave from Johns Hopkins.

I went to the first meeting of Math 3 with a certain amount of fear and trembling. Would I be up to it? Prof. Zariski opened up his black ring binder notebook from which he lectured in a heavy European accent. Immediately he put some matrices on the blackboard. Thank heavens for my reading in the Carnegie Branch Library on Twenty-Third Street! I decided I would be able to handle the course. I was 17 at the time. Zariski was 41. To a student, a professor appears as a fixed figure; a snapshot. Zariski was severe, careful, solid, unbending, demanding, stimulating. In my camera lens, Zariski appeared as the Great Stone Face.

I found out only recently with the publication of Carol Parikh's substantial, if somewhat plain biography, *The Unreal Life of Oscar Zariski*, (1991) that Zariski was born in the town of Kobrin, about fifty miles east of Warsaw and now in Byelorussiya. He was at the University of Kiev from 1918 to 1920. Thence to Rome where he obtained his Ph.D. in 1924. There he came in contact with the famous school of nineteenth- and twentieth-century Italian geometers that included Castelnuovo, Enriques, Severi and Segre. He had already moved strongly into modern abstract algebraic formulations of algebraic geometry (which did not much interest me in my later career. I thought the balance between the visual and the symbolic was upset) but which Castelnuovo predicted correctly would be a shot in the arm for the enervated Italian development.

Zariski was at Johns Hopkins from 1927 to 1937. He received a permanent appointment at Harvard in 1948 and remained there until his retirement in 1968. The decade from 1935 to 1945 was, paradoxically, the period of Zariski's life of the greatest agony and personal suffering, and simultaneously the period of his greatest mathematical creativity. The sweep of Hitler's armies into Poland put his family remaining in Kobrin at risk, and after he taught our Math 3 class, he would later go home and hang on the radio to hear of the progress east, kilometer by kilometer, of the Nazi hordes. His mother and older brother were swallowed up by the Holocaust, a loss to which he was never reconciled psychologically.

As part of an anodyne process of consigning the past to oblivion, Zariski abandoned his Jewish background, mutated politically from Communist to Republican, and attempted to lose himself among the old New Englanders. Mathematics can also be a powerful anodyne: anyone who has ever drunk deep at the springs of mathematics knows that they are often located in an alternate world far, far away from the world of ordinary pleasures and pains.

I was in class with Zariski three times a week for a full year. I never really got to know him. But I loved his course and he gave me an A. He marched the class through homogeneous coordinates, harmonic division, cross ratio, projective coordinates, the geometry of the complex plane. Ah yes, there I met up with the two imaginary points at infinity I and J (called by some aficionados of projective geometry Isaac and Jacob). These points possessed a doubly absurd reality, way off at infinity (wherever that was) and bearing imaginary addresses (whatever that meant). Whatever they were epistemologically, they were points with which one could argue consistently and productively. Though they went beyond and transcended visual imagination, they grabbed me immediately. A conic was a circle if and only if it passed through the points I and J. Here was mystery; and the three-way tension between what was visually imaginable, what was symbolic and what was virtual had a sensual quality.

Nonetheless, I kept wondering if would we ever get to Pappus.

We got to him fairly late in the course, when we came to the projective theory of conics. With all the projective machinery available, we proved Pascal's Theorem, which Blaise Pascal in 1640, at the age of 16(!), had discovered and proved, and which was a generalization of Pappus. Pascal had discovered that if you start with a conic section (ellipse, hyperbola, parabola) and take six points at random on it and cross connect the points as in Pappus, then the three points of intersection (as in Pappus) are always collinear.

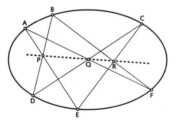

Pascal's Theorem.

Pascal implies Pappus. Why? Simply because the initial two straight lines of Pappus can be considered a degenerate conic section.

Having arrived in Math 3 at Pappus through Pascal, my interest in projective geometry was slaked and slackened. But not, as we shall see, my interest in Pappus. As regards my teacher, Oscar Zariski, he was even then developing into a formidable innovator of worldwide renown in the field of algebraic geometry.

Although he was famous in the universe of mathematics, and very prominent on the American mathematical scene, after the last class of Math 3 and the final exam, he never crossed my path again. But my interest in Pappus abided. In the post-Pascal period, Pascal's theorem (often called the "Mystic Hexagram," and to be distinguished from Pascal's triangle of binomial coefficients) grew into a veritable cottage industry with its devotees and an immense literature of its own. The theorem was extended, generalized, deepened to a fare-thee-well. Pascal is himself reputed to have derived four hundred theorems from his one theorem, but his writings on the topic have been lost.

As an example of what is around: six given points will, when taken in some order, determine sixty different hexagrams. If these six points lie on a conic then sixty different Pascal lines will be determined. These sixty lines fall into twenty groups of three, each group of three lines passing through a common point. These twenty points lie by fours on fifteen lines, three of the lines going through each point. If all this were drawn out in a single diagram (and it has been), it would appear as an absolute visual madhouse of lines and points with some order here and there.

But there is even more to the Pappus Theorem. It is often regarded as one of the first theorems of projective geometry, in advance of the subject by fifteen hundred years. When projective geometry, getting off the ground in the seventeenth and eighteenth centuries, itself became abstracted and axiomatized, all geometry became totally symbolic and moved out of the visual and the usual two- and three-dimensional spaces of physical appearance and experience. In such a formal environment, there could be not just one standard paradigmatic geometry, but as many as one pleased, all distinct and all logically consistent within themselves but mutually contradictory. In particular, and very strangely, there could be geometries that had only a finite number of points and lines.

Under the program of axiomatization and generalization, Pappus's theorem changed its status from a theorem to an axiom. An abstract, or simply, a more general kind of projective geometry can be built up as a set of homogeneous coordinates, (a,b,c) where the components a,b,c are not restricted to be numbers, but can be elements of any algebraic system in which one can add, subtract, multiply and divide (a so-called division ring). Now it has been established that if the multiplication in the division ring is commutative, i.e., if $fg = gf$ for all elements f and g, then Pappus's theorem is valid, and vice versa. Thus, if it is convenient or if it has aesthetic appeal, Pappus can be taken as one of the basic axioms of an abstract projective geometry. We can speak now of Pappian planes and non-Pappian planes. (Here, in this language, we come a wee bit closer to where Zariski resided professionally.)

The interchangeability that exists between assumption and conclusion, i.e., between axiom and theorem, (and Pappus is merely one instance out of many) totally alters the notion of an axiom as a self-evident basis on which a mathematical theory must be founded. Pushed to the limit, it can display mathematical truth as simply the progression from one statement to another, the progression being carried out according to the various formal "if-then" laws of logic. This is a view of mathematics that I have always found inadequate. I have found that I never understand a piece of mathematics thoroughly until I have digested it in my own way, on my own terms. My way may turn out not to be the easiest or the cleverest or the current way, but as in the old song, it is my way.

For some unknown reason, I was never satisfied with the proofs of Pappus I knew about, and for a long time I wondered whether it would be possible to muck through with my own proof, simply by using straightforward coordinate (Cartesian) geometry. The method of Descartes (1637) is frequently regarded by students and teachers as a "machine" into which one feeds the hypotheses of a certain geometric situation, and which is guaranteed to grind out the desired conclusions given sufficient patience on the part of the problem solver. However, it is no denigration of Descartes to assert what also has long been known: that many elementary situations give rise to impossibly long and tedious computations, and hence Descartes' "universal method" which replaces brains by brawn, founders upon the rock of limited human patience and endurance. Ways around are then sought; these include clever coordinate systems, special transformations, determinants, abridged notations, special devices and tricks, etc.

A preliminary computation showed me that if I went this road, with no particular cleverness employed, then in no time flat, I would be lost in a welter of symbols (monomials), hundreds of them, perhaps more. This was no work for mortals, at least not for me.

My Cartesian approach using "artificial stupidity" had to wait many years until digital computers became available. And beyond that, it had to wait until formal algebraic software became available for them. With such software and with sufficient memory, I would be able to get through.

The day came for me in 1969. FORMAC, an early symbolic computation package, had become available. Elsie Cerutti, a top-notch programmer, was around. I outlined the program, using a very slight amount of intelligence to reduce the complexity, and "pushed the button" on the IBM 360/50 (of blessed memory) with only 256 K bytes of core storage, and after 4.52 minutes of execution time, received the print out: $DE = 0$.

The computer had simplified 246 terms inherent in a critical determinant DE and they all cancelled out (or added up) to zero. Therefore Q.E.D. for Pappus.

Fooling around with and interpreting other printouts from Elsie's program, I found it was a simple matter to derive many other theorems relating to the Pappus configuration, theorems that were probably not in the recorded literature. Pascal's mystic hexagram contains worlds within it.

This success with computer-assisted proofs did not yet saturate my enthusiasm for Pappus. It occurred to me that to get a computer-assisted proof of Pappus, one did not have to have available a program that did formal algebra. It would suffice (in a certain sense) if one worked with standard numerical software, provided one were able to start with six points that were algebraically independent over the reals. How to do this, and what algebraic independence could mean with finite precision numerical software, were the subjects of several papers I wrote with Prof. John Rowland of the University of Wyoming.

With all these experiences at hand, philosophical questions inevitably raised their slippery metaphysical heads. What is the distinction between a "transcendental element" and a pure symbol? Does a computer-assisted proof have less validity than a totally hand-crafted proof? Is any computation (by humans or by a machine) essentially a theorem? Is there an analogy between a computer-assisted proof and a physical experiment? If theorems can be spewed out of a computer like doughnuts from a doughnut machine, what are the criteria for distinguishing which theorems are and which are not of importance? If theorems are a dime a dozen, what can one say about the "platonic inevitability" of any particular one? Can definition and proof really dispel a mystery? (Speculations on these and other questions grew in my mind into a philosophy of mathematics that is elaborated in the book *The Mathematical Experience* that I co-authored with Reuben Hersh.)

Computer-aided mathematical proofs in geometry are now all over the place, relying not on artificial stupidity, but on sophisticated constructs such as

Reuben Hersh and Phil Davis.

Groebner bases, and the interest in such things has popped up a metalevel, from the individual theorems (of which there are a super-plethora) to the methodology of the theorem proving. One of my literary friends has told me that one of the hallmarks of post modernism is its self-referential quality. Here, just as on TV we have soap operas whose sole subject is the making of soap operas, we see this quality in mathematics.

And have I, at the age of three-quarters of a century, exhausted my interest in simple Pappus? Who knows? *On revient toujours à son premier amour.*

Post Script: Perhaps even more fundamental to axiomatic projective geometry than Pappus, but of greater visual complexity, is the famous theorem of Desargues: "Two triangles that are in perspective from a point are in perspective from a line."

In abstractions, as with Pappus, this becomes an axiom and the relationship between these two axioms has been studied extensively. Of this historical development, Gian-Carlo Rota has written perceptively,

> The *value* of Desargues' theorem and the *reason* why the statement of this theorem has survived through the centuries, while other equally striking geometric theorems have been forgotten, is in the realization that Desargues' (as well as Pappus's) theorem opened a horizon of possibilities that relate geometry and algebra in unexpected ways... What an axiomatic presentation of a piece of mathematics *conceals* is at least as relevant to the understanding of mathematics as what an axiomatic presentation *pretends* to state.

Mathematics and Showbiz:
Shall Never the Twain Meet?

Summer of 1941. I had just finished my sophomore year at college. While the Soviet Armies were slowing down Hitler's hordes considerably, the outcome was still problematic. Shostakovitch had completed his *Leningrad Symphony*. After providing Lend-Lease, Bundles for Britain, and other such gestures, America was moving slowly towards its entrance into World War II. We were all registered for the Draft.

And I was working in the Tavern of Tamiment, an adult camp in the Pocono Mountains near Bushkill, Pennsylvania, selling candy, ice cream, cigarettes, ping-pong balls, and minor drugstore sundries. For my summer's work I got paid room (in a bunk house), board (staff dining room below that of the guests), and a cash payment of $100 which amounted to half a semester's tuition at Harvard. I got the job through pull. My brother's college classmate's uncle ran the dining facilities at Tamiment. That's the way connections work.

Camp Tamiment had a relationship to the American socialist movement and to the Rand School of Social Science (1906–1956) in Manhattan. Despite its socialist origin and leanings, by the time I got to Tamiment, it had become quite bourgeois and the camp operated as a high-class establishment and charged accordingly. For their money, the guests got good bunks, excellent food, swimming and boating facilities, tennis, golf, and "calisthenics at the waterfront." There was dancing to a resident dance band, there were lectures by well-known personalities, usually on social themes, there were movies, and the crowning glory of the place was the "Tamiment Playhouse" (1921–1960): live entertainment, generally musical revues. A new revue each week! Think of that.

Phil Davis and friend, 1940.

To get there, I was instructed to pick up the bus to the camp at the Rand School on East Fifteenth Street. For relaxation, I had packed in my bag *Advanced Calculus* by Frederick Woods, professor of mathematics at MIT, meaning to work my way through it. Near the Rand School there was a secondhand book store displaying its wares on the street and at the last moment, before jumping on the bus, I bought a novel by G. B. (Gladys Berwin) Stern for five cents. This occurred at the very height of my novel-reading days, but I hated the book.

I arrived at Tamiment just two days after the camp had dedicated its newly constructed theater building next to the Tavern. This was no summer-theatre-in-the-barn, with hay and chickens in the loft. This was professional stuff with professional aspirations and, as it turned out, with professional Broadway and Hollywood accomplishments. Led by Max Liebman, who was writer, director, and impresario all rolled into one, and who later, in the infancy of television, put together the successful "Show of Shows," the Tamiment Playhouse was the seedbed of much young new entertainment talent.

I worked there in the summers of 1941 and 1942, and in those years, working beside me (so to speak) as entertainers, were Betty Garrett, Jerome

Robbins, Carol Channing, Jules Munshin, Imogene Coca, and Anita Alvarez, all of whom graduated from Tamiment to the national bright lights. In previous years, Danny Kaye and his wife Sylvia Fine who was pianist, composer and lyricist had been there. In later years, Carol Burnett and Woody Allen also put in time at Tamiment.

I had to work practically every evening, and when the show broke and the guests lined up at the Tavern counter, I and my fellow bottle openers really had to hustle. The afternoons were fairly quiet and in between customers, I would sit behind the counter, working away at advanced calculus. Green's theorem, Stokes's theorem, the gradient, the divergence and the curl of vector analysis, even today put me in mind of Pola-Cola, a local competitor of Coke, whose bottler in Stroudsburg, Pennsylvania, insisted was every bit as good and at half the price.

Sometime in August, the word went around the Tamiment staff that in a few days Alexander Kerensky would be visiting and giving a talk. Did I know who Kerensky was? Well, just barely. I knew that he was a pre-Lenin, pre-Trotsky, pre-Stalin political figure in the Soviet Union. He was mentioned in articles and in the history books, and they said that whatever it was that he tried to do, it had failed. Kerensky soon lost out and later went into exile.

In the middle of a quiet afternoon, while I was reading Woods's *Advanced Calculus* behind the counter, Alexander Kerensky, surrounded by a few Tamiment bigwigs, came up to my counter. He asked me what cigars I would recommend. I told him he could have anything from a ten-cent Philly to a fifty-cent Corona Belvedere that came in its own aluminum foil tube. Kerensky, like the good middle-of-the-roader that he was, settled for a La Primadora at seventeen cents. He then asked me what I was reading. I told him. He nodded his head in approval. (Had he studied calculus? Did he know Green's Theorem, or, as the Russians call it, Ostrogradsky's Theorem?)

And then there was the day when Danny Kaye, already a tremendous star for his role in "Lady in the Dark," visited his Alma Mater. He sat in a table in the Tavern, surrounded by a retinue of retainers and a circle of fawning admirers, doing his shtik and convulsing us all with laughter.

Although I saw the entertainment staff regularly when they came into the tavern to buy ice cream or suntan lotion, and although one of them, I think it might have been Carol Channing, used to call me (quite inappropriately, I thought) "The Sphinx," for my studying calculus behind the counter, the only one I really got to know was Bob Burton. Burton was an actor, a song-and-dance man, and a skit writer for revues. He was married to Coca. Burton used to come in regularly for cigarettes and we would talk at length. Not about calculus, heaven forbid, but about my other love: the short story.

I had in mind to write a skit and show it to Burton. My skit would be of the burlesque type (what else did I know?) and I outlined it to him. He said: go ahead. Did I really think I had what it takes to be a writer for Broadway? Was this a fork in my road through life? I never wrote the skit. The pull of Green's theorem was too great.

Philipp Frank

He was a one *l* and a three *p* Philipp. This endeared him to me immediately. I wish I had a dime for every time people have asked me whether there are one or two *l*'s in my first name.

He was a short, stocky man, and shuffled as he walked. I think he had had an accident. I remember him as having a walrus moustache, like Justice Oliver Wendell Holmes; the sort of moustache whose ends become yellow from tobacco or snuff. He spoke haltingly and with a thick accent. Fair enough, he'd been in the country only since about 1939. I loved his planet Jupiter. He pronounced it thus: Der Yuppiter. I remember how, to make a philosophical point, he ripped off a strip of newspaper, held it up, and let it flutter to the floor in the drafty lecture room. "We can never predict with great accuracy where the paper will end up on the floor."

His name was Philipp Frank (1884–1966). A student of the famous Ludwig Boltzmann at the University of Vienna, from 1912 to 1938 he had been professor of physics at the German University in Prague. I knew that he was the Frank of the famous book Frank-von Mises: *Die Differential- und Intgralgleichungen der Mechanik und Physik*, itself an update of the earlier book on the topic by Riemann-Weber.

Frank was Einstein's successor in Prague. He was exiled by the Nazi takeover of Czechoslovakia. In 1942 he was a lecturer at Harvard and offered a course in the history and philosophy of science. The scuttlebutt among my fellow undergraduates was that it was a "gut course—but good." Students who were not interested in science took it as an easy way of fulfilling their distribution requirements in science. I took it. I enjoyed it.

Philipp Frank.

I loved the man. His students called him Philipp as a mark of love and respect and not, as in these days, as a pseudo-democratization of rank. My interest in the philosophy of science and mathematics, my concern also with the relationship between mathematics and society derives from my having sat at Philipp Frank's feet.

Not only did I learn about Kepler and Newton and Einstein, and his fellow scientist/philosopher Percy Bridgman, but I learned about Jacques Maritain, Teilhard de Chardin and a host of other humanists he admired. I remember particularly that I first heard of William James's "tough-minded" personalities (materialistic: just give me the facts, please) and "tender-minded" (idealistic: let's make a beautiful system) from the lips of Philipp Frank. The amount of material he presented in class was modest. He did not try to stuff us with information. Less is more. He got us to think for ourselves. And that's what a good teacher is able to do.

Yes, Frank got his students and those who read him to think for themselves. Paul Feyerabend (1924–1994, a brilliant Austro-American philosopher of science with maverick tendencies) remarked in his autobiography *Killing Time* that Frank argued that the Aristotelian objections against Copernicus agreed with empiricism while Galileo's law of inertia did not. As in other cases, this remark lay dormant in my mind for years; then it started festering.

I recall sitting in the classroom in Jefferson Laboratory where Frank held forth and hearing just those same words and being shook up by them.

In those days, the vast stacks of Widener Library were closed to undergraduates. However, if you were doing an senior honors thesis, your tutor could get you a stack permit. I remember how "professional" I felt when Prof. Lynn Loomis, my tutor, took me by the hand, so to speak, and showed me how the stacks and the carrels worked. More importantly in terms of professional development, he introduced me to the immense collection of volumes known as the *Fortschritte der Mathematischen Wissenschaften*, a reviewing journal for all important mathematical articles going back to the 1880s. Mathematical German never posed any problems for me. German mathematics was another matter.

I would be sitting at a carrel in Widener Library around 8:00 PM. I would hear a shuffling noise toward the entrance of the stacks. It must be Philipp come to check out a reference. He approached my carrel. I hailed him. He stopped. He told me to sit, not to get up. He talked and talked about the philosophy of physics. I put in a word once in a while. That semester this scene was repeated numerous times.

Philipp, a logical positivist, had been a member of the famous Vienna Circle—the *Wiener Kreis*. Even as I was learning rather more about logical positivism (or logical empiricism as it was sometimes called) than he presented by indirection in class, I was beginning to formulate my own philosophy.

The characteristic theses of the Vienna Circle were developed mainly by Moritz Schlick, a philosopher; Hans Hahn, a mathematician; Otto Neurath, a political economist; and Rudolf Carnap, a logician. I cannot here go into the details of logical positivism; in brief, its philosophy asserted the primacy of observation in arriving at the truth of statements of fact, and it ruled out metaphysical and subjective arguments as being meaningless nonsense. Philipp Frank had this to say in his book *Between Physics and Philosophy*:

> The thesis (of the Vienna Circle) that has perhaps become best known is that a proposition has a meaning only if it states the means for its verification. From this follows the "meaninglessness" of all metaphysical propositions, by which is meant, of course, "meaningless" for science, not the denial of any influence upon human life.

Despite Philipp's warning that the Circle applied the word "meaningless" only in the context of science, most people assumed (well, at least I assumed) that logical positivism ruled out metaphysics, *tout court*; being, even at a young age, soft on metaphysics, I broke with logical positivism without having

Phil Davis and David Park.

really drunk deep of it. My skeptical genes balked at systems that lay down what something is or ought to be. Mathematical systems may be the exception, and even there I'm not certain.

For another student's view of Philipp, I quote from a letter of David Park, who is a distinguished quantum physicist:

> There was a time during the war (World War II) when I saw quite a lot of the Franks. Harlow Shapley (famous Harvard astronomer) had invited Philipp to write a book, to be called *The Relativistic Universe* for a series he was editing. For some reason Philipp thought I would be a good junior partner. We agreed that tensor calculus wasn't the way to go and I produced a few chapters of an alterative approach which I now realize wasn't right for the occasion.
>
> One of the things I learned about Philipp was that he was incapable of saying something was wrong even if it was. His strongest word was "awkward," but he didn't even use that word on me, and so without any firm criticism, which I would have welcomed and profited from, the project drifted. After a while I saw nothing was going to happen and turned to other pursuits.
>
> This was all before I started graduate school. In class, he asked for a paper from each student, but we were to write on whatever pleased us. I don't know what bromide I turned out, but

Clara (Claiborne Park) told me that a friend of hers had recycled a perfectly good paper on John Crowe Ransom and the Southern Agrarian Movement. Philipp thought it was fine. I remember most clearly Philipp's extraordinary gentleness and kindness.

The Nazi period in central Europe spelled the end of the *Wiener Kreis*. Schlick was murdered (non-political) on the steps of the University. Many of the members of the circle were fortunate enough to get to the United States and to set up their main shop at the University of Chicago, where from time to time they put out a volume of *The International Encyclopedia of Unified Science*.

The Lecture Room in Vienna

A few years ago I was in Vienna and wanted to see the room where the Vienna Circle met. Karl Nagelmacher, who teaches at the University of Vienna, met me at my hotel and we took the tram to the University. We walked down the corridor of the Mathematics Institute on *Strudlhofgasse*. A sign over one of the doorways that remained from the days of the Empire read: "*K.u.K. Mathematisches Institut.*" Karl found and entered a certain classroom.

"This is it. The Vienna Circle met here throughout the 1920s and into the mid '30s. It's a classroom now that 'belongs' to The Institute for Meteorology and Geophysics. Every now and then, a club called the '*Verein zur Foerderung Wissenschaftlicher Weltauffassung*' (Society for the Promotion of the Scientific World View) meets here and sponsors talks, books, and research projects."

I entered the classroom and saw "ghosts." On the wall, there were posters and photographs relating to the Circle, its activities and personalities. I spotted pictures of Moritz Schlick, Friedrich Waissmann, and Rudolf Carnap. Then I spotted a picture of Philipp Frank, my teacher of a half-century ago, a young man when the picture was taken; not so young in the years I knew him. This, then, was the room where throughout the 1920s and into the mid 1930s, the members of the Circle met and carved out a new philosophy of science: logical positivism.

The origins of the Circle, around 1910, are to be located in the positivism of the physicist-philosopher Ernst Mach (e.g., Mach 3 in aeronautics and in science fiction; same guy). The new logic was developed by Frege, Schroeder, Russell, Whitehead, Wittgenstein, and numerous others. The Circle was sympathetic to the semiotic ideas of Charles Sanders Peirce, the pragmatism of William James, and the operationalism of Percy Bridgman, Americans all three.

Phil Davis in the Vienna Circle Lecture Room.

A frequent, but not universally held, view is that philosophy must always follow experience; it cannot really compel experience. Operationalism, in the mind of physics Nobelist Percy Bridgman, says, briefly, that all concepts introduced in physics must be accompanied by precise descriptions of the operations (often measurements) that are required to give meaning to the concepts. Logical positivism is sympathetic to operationalism.

How has operationalism fared since Bridgman published his book *The Logic of Modern Physics* in 1927? I asked David Park and received his answer:

> Nobody mentions it any more, that's certain. But I think it forms part of our thinking. There are by now plenty of physical quantities that we use all the time and are not capable of measuring. (The energy of an atomic system, for example). Most of the things we talk about can be measured, at least in principle, and good physicists tend to think of them that way rather than as quantities that emerge from a computation.
>
> Here is a speculation. In Bridgman's day, atomic and particle physics were very little developed. By now we are conscious that, loosely speaking, every transformation that leaves the [mathematical] description of a physical system invariant has associ-

ated with it a property. That property is given a name, but in many cases it is so far from the laboratory and, correspondingly, so enmeshed in the quantum-mechanical theory that one would be puzzled how to make a convincing measurement.

Despite the fact that I did not subscribe to the program of logical positivism in its narrow sense, I found support and sustenance in its wider, humanistic goals. Here are a few of the positions I found attractive in Philipp Frank's 1957 book *The Philosophy of Science*. After quoting Ortega y Gasset to the effect that the average scientist is a "learned ignoramus, which is a very serious matter as it implies that he is a person that is ignorant not in the fashion of the ignorant man, but with all the petulance of one who is learned," Frank dedicates his book to breaking down the existing barriers between science, metaphysics and philosophy, since "the great majority of today's physicists have been successfully trained to keep their special fields as separate from philosophy as they can."

Originally, there was a strong link in the minds of scientists between science and philosophy. But then, technology became scientific and "the union of science and technology was responsible for the separation of science and philosophy."

The "tough-minded" scientist was born. Does the theory describe the facts? Does it have predictive value? This is all the tough-minded scientist wants, and away with irrelevant and obfuscating speculations that bring in metaphysics or philosophy or religion, politics, or ethics. Away with the whole human element of the thing. However, this element cannot be spirited away so easily, and Frank shows how it turns up time and again to embarrass and put on the spot scientists who have no opinion or who may be in the grip of some childhood philosophy.

In its day, logical positivism was greatly influential and not only did mathematicians and physicists come under its sway, but economists and sociologists as well. By the 1950s, the opinions of the logical positivists were themselves beginning to turn a corner in a direction that was consonant with my own thoughts.

It was under the imprint of the positivistic *International Encyclopedia* that the tremendously influential 1962 work of Thomas Kuhn *The Structure of Scientific Revolutions* was published. I think, but I am not sure, that Kuhn heard Frank's lectures as an undergraduate. Richard Rorty recently wrote: "Most of us philosophy professors now look back on logical positivism with some embarrassment, as one looks back on one's loutishness as a teenager." (The *Wilson Quarterly*, Winter, 1998, p. 33)

Despite my early break with positivism, I revered Frank and realized how important a person he'd been in my intellectual development. Was this class-room on *Strudelhofgasse* then, one of the holy spots of my life? What does one do at the holy spots? Should I put on a hat and take off my shoes as Moses did before the burning bush? I gave Karl my camera, and asked him to take a picture of me standing up against the picture of Philipp Frank. That would have to suffice, I thought, and we walked back toward the main lecture hall.

Louis Finkelstein

There is a saying among mathematicians that mathematical talent is often passed down from father to son-in-law, and there are numerous historic cases to back this up. In my instance, I learned much from my father-in-law, Louis Finkelstein (1895–1991), but it wasn't mathematics. He himself knew very little mathematics, and every once in a while to my great annoyance he would trot out Whitehead and Russell as the culmination of mathematical achievement in our day. Perhaps he inherited this opinion from his teacher at City College of New York, Morris Raphael Cohen (1880–1947), who himself was a logician, a positivist, schooled in the Russell tradition. Cohen wrote that he considered himself a person whose main contribution was to have cleaned out the Augean Stables of Thought.

I first met LF in 1942 when I was dating his daughter, Hadassah. He was then the President (later: Chancellor) of the Jewish Theological Seminary of America, located at Broadway and 122nd Street, Manhattan. I could see at once that LF had a robust sense of humor, a match for my own, though mine is rather more ironic, and I, who would walk a mile for a good punch line, said to myself that if only for this reason, LF was a possible father-in-law.

In those early days of our relationship, I doubt whether he cared very much about the state of my wit, but only whether I was old enough, whether my intentions towards his daughter were honorable, whether I was stable, whether I could make a living through mathematics. And those questions, I recall, I could only answer or parry with my own question of whether, being draft liable, I would survive the war that the United States had just entered.

I think that LF saw quite early that I was not an ignoramus about Jewish matters. My father's father, who came from the Ukraine, and was naturalized as

Louis Finkelstein, 1960 (Photo by Joseph Costa).

an American citizen in 1895, was orthodox in his practices. My parents moved over to Conservative Judaism (of which community LF became the titular head a few years later). Sometime in the early 1920s my father bought a copy of the *Jewish Encyclopaedia* in twelve volumes, and as a child I cut my reading teeth rummaging at random through its pages.

If my knowledge of Jewish history, liturgy, and ritual were passable, my grades in personal ritual observances were D plusses. My non- (occasionally anti-) ritualism didn't seem to concern LF too much. He never spoke to me about it; never once, over the years, did he try to "convert" me to a more intense level of personal religious observance. Early on, I arrived at a *modus vivendi* that consisted of rendering unto Louis the things that were his and maintaining my own comfortable level of ritual slackness. I maintained my anti-Platonic philosophical beliefs, and for all that I was clearly headed for a career as a mathematician, I was not entirely a rationalist.

Yes, LF thought that I was more than a bit of a mystic and an irrationalist. I thought of myself as a philosophic skeptic, a satirist but not a scoffer. I have never felt embarrassed by the words of Psalm I, Verse 1 in which one is cautioned "not to sit in the presence of scoffers." For all that I have tried to teach my students to consider deeply the reasons for their beliefs, I have rarely mocked.

At the time of my wedding to his daughter Hadassah, LF was in full limelight both as regards the American Jewish community and the secular community. In 1951, his picture was on the cover of *Time Magazine*. In 1956, he was asked to participate in the inaugural ceremony for President Dwight Eisenhower and Vice President Richard Nixon. In 1961, he was named by President John F. Kennedy as one of several (religious) American representatives to the installation of Pope Paul VI in the Vatican.

Around 1940, LF inaugurated an annual conference on science, philosophy and religion whose objective was to pull together these areas. Science and religion, in particular, had been at each others' throats for many years; the scientists of LF's generation were for the most part atheistic or agnostic, and it took rather a bit of persistence and courage on LF's part to persuade them to sit down in the same room with theologians. I suggested to LF one year that he might invite Philipp Frank to speak and that worked out.

I learned many things from my father-in-law. The meaning of scholarship, its humanistic values, and—this is very important—the very motions through which one pursues it. Though I had been interested in philosophy since undergraduate days when I sat at the feet of Philipp Frank, my philosophic interest intensified considerably as a result of my occasional discussions with LF. (He had a Ph.D. in philosophy from Columbia.)

Through my father-in-law, I learned of the medieval Jewish philosopher Saadia (882–942), about whom he had written a book. I found Saadia's exceedingly dry and abstract approach to God very axiomatic and scholastic; not entirely sympathetic to me but eye-opening, and some years later, I put in a long quotation from Saadia as an instance of how abstraction, whether philosophical or mathematical, when carried to its limits, can be dehumanizing.

It was through association with LF that my interest in general history intensified considerably—it was already present from high school days but limited to the history of mathematics. One scholar whom LF admired greatly was Harry Wolfson, Professor of Philosophy at Harvard, a specialist in Philo of Alexandria and Spinoza, and when I returned to graduate school in the fall of 1946 to do a Ph.D. in mathematics, I took time out to listen to Wolfson's historical lectures. The Wolfson experience fed back into my anti-Platonism in that I cannot really conceive of knowledge except in a historical context. Much later, I was able to exploit this position as part of a "new wave" (anti-foundationist) movement in the philosophy of mathematics.

I learned from LF how, in a public speech, to show yourself as an individual human being. The speaker on the podium operates largely as an ex-officio abstraction. When, therefore, the speaker reveals a bit of his personal life, he becomes Everyman, and the audience will identify strongly with him.

Louis Finkelstein with Phil and Hadassah Finkelstein Davis, 1962.

LF knew how to do this marvelously well, but never fell into the trap of becoming maudlin as Richard Nixon was in his "Checkers" speech.

Toward the end of his long life, as things became increasingly difficult for him physically, he said: "I thank God every day that I am dying from the bottom up and not from the head down."

Part III

World War II

First Real Mathematical Job: The A.S.T.P.

June 1943. Commencement. A good many of my classmates, as well as the faculty, had already volunteered or had been drafted into military service. Some deferments were allowed on the basis of majoring in science. The class that remained had been reduced to a shadow. Many would come back after the war was over, benefiting from the GI Bill; a few would fall and their names would be preserved in marble on the wall of Appleton Chapel.

At that time, the war seemed interminable. In Europe it was already four years old. Time and the normal course of life stopped, particularly for the young. The war was fueled, possibly driven, by science and technology, and the nation developed more and more of its resources for a long conflict. Young mathematics faculty members left Harvard and took positions in groups doing such things as radar, cryptography, atomic energy, and operations research.

As part of girding up the national loins, someone had the idea of teaching soldiers and sailors a bit more science than they would normally have learned. One of the programs was known as the ASTP (Army Specialized Training Program).

In the summer of 1943, the ASTP program was in full swing at Harvard, and I and a number of my fellow science majors with brand new bachelors' diplomas in our hands were swept into teaching enlisted soldiers, many of whom were older than ourselves. I'm not sure, even at this late date, just what the point of the program was, but that didn't matter. Nothing mattered much and the draft boards might be kept off one's back for a few more months.

The summer was hot, the students and the young faculty were restless. Our future was moot. And there we were, we happy few, so to speak, teaching poorly prepared classes a bit of algebra, a bit of trigonometry, and with some luck a bit of calculus.

I don't know when the idea of having students give grades to their courses and their teachers became popular, but it was already in place when I first taught. In clearing out some boxes the other day, I came across a yellowing four-page document entitled "General Criticism of ASTP by One Informant." I reproduce here its final lines:

Summary of Suggestions

- Less acceleration
- More time for studying
- Better conditions for studying (i.e. fewer men to a room)
- Axe the military: either have a military camp or a school. They don't go well together.
- More individual instruction (or opportunity for learning) in Physics and Chem.
- I think math is okay.

Langley Field, Virginia

My first research job was at the NACA (predecessor of NASA) at Langley Field, Virginia, in the spring of 1944. For a while, my wife and I lived in a trailer park in Newport News erected to house workers in the vast shipyards nearby. Newport News contained a turbulent mixture of soldiers, sailors, airmen, their wives, husbands, children and sweethearts, from all over the country. Our trailer was adjacent to a camp that housed Italian POWs.

The NACA laboratory was staffed by old-timers in the business plus many young draftees, hotshots in mathematics and physics who were assigned to this work as their military service. In due course, I received my "Greetings" from the Selective Service System

August 19, 1944

ORDER TO REPORT FOR INDUCTION

The President of the United States,

To   Philip Davis

GREETING: Etc. etc.,

and was inducted into the United States Air Force and assigned to the reserve.

I worked in the Aircraft Loads Division, which carried out experiments and analysis of the aerodynamic loading of the wings and tails of fighter air-

Phil Davis and Phil Rabinowitz working through a numerical strategy in the first generation of computers.

craft. We were particularly interested in stresses under maneuvers: dives, pull-outs, fishtailing, evasive actions.

We obtained our data from tubes inserted in small holes drilled along a cross section of a wing or a tail. The tubes were connected to air pressure recorders. On the basis of these records we would construct the profile of air pressure across the section as the aircraft performed a variety of maneuvers. From the pressure profile, we computed two very important aerodynamic quantities related to wing strength and aircraft stability: the lift coefficient and the coefficient of pitching moment. The pressures at the various stations across the wing or pail were plotted up. These isolated values were then faired in with a French curve to give a complete pressure profile. The area under the pressure curve was essentially the lift coefficient and the horizontal moment of the area was essentially the pitching coefficient. Finally, as a last step, a special instrument called a planimeter was used to obtain the graphical areas. After tracing the stylus of this instrument around the boundary of the area, we could read the area off on a dial. If the planimeter was a real fancy one, and ours was, it would have two dials, the first giving the area and the second the horizontal moment.

Our super-duper planimeter was in the permanent and personal care of an engineer I shall call Swindells Royce—Dell, for short. Dell treated the shining precision instrument with the care of a mother hen, or, better still, with the care of a jeweller in charge of a fancy Swiss watch. He was always polishing it, oiling it carefully, locking it up when it was not in use. And it was well that he did so; it was a sensitive, temperamental and rare piece of equipment. The prototype was German, the Germans being, in those days, the finest instrument makers. But German instruments being unavailable during the war, the machinists at NACA took apart one they had and made ten copies at a cost, I was told, of $5000 apiece. This was at a time when a car, if you could buy one, would have cost $600 to $800.

It occurred to me, as a fledgling mathematician, that the whole expenditure was unnecessary; that a little bit of carefully laid out arithmetic (numerical integration) could have replaced both the drawing of the diagrams and the subsequent planimetry, and would have yielded answers that were just as accurate. But planimetry was the way it had been done, and it was the way it was going to be done, and I sensed that I had better shut up about it. And so I pushed the stylus around many a pressure diagram before the war had ended. This experience led me to an intense interest in the theoretical aspects of approximate numerical integration and years later to a book *Methods of Numerical Integration* written with Phil Rabinowitz of the Weizmann Institute, Rehovot, Israel, as co-author.

Part IV

Graduate Years

The Clock Runs Again:
Graduate School

World War II was over. VE Day (Victory in Europe) was May 8, 1945. VJ Day (Victory in Japan) was August 14, 1945. Demobilization of the U.S. military forces set in rapidly. Rationing of gas and foodstuffs was terminated. Many of the young scientific inductees at NACA left Langley Field and went on to become some of the leading scientists and technologists in the country. By September 1946, I was released from the Air Force Reserve and reentered Harvard Graduate School in the department of mathematics.

Life's clock, which had absolutely stopped during the war, was running once again and the years were full of ferment. The universities of the country were filled with returning veterans studying under the GI bill. Some were completing their interrupted undergraduate education. Many others were pursuing graduate studies. The almost standard dress of Harvard students in those days was olive drab pants of army issue and dark brown leather loafers, topped off by a tweed jacket.

The faculty of the Department of Mathematics included Garrett Birkhoff, Joseph L. Walsh, Hassler Whitney, Lynn Loomis, Lars Ahlfors, George Mackey, David Widder, and Saunders MacLane. Among the younger faculty, I recall Lowell Schoenfeld, Bob James, Gerhard Hochschild, and Andy Gleason. There were distinguished visitors to the department; for example, Henri Cartan from France and Arne Buerling from Sweden. Those from Central Europe or England struck me—no surprise—as a bit worn out physically and poorly clad.

A few buildings away, in the Department of Applied Science, were Richard von Mises, Stefan Bergman, and Max Schiffer. Just adjacent was (Cmdr.) Howard Aiken's computer laboratory, where the relay clunker known as the Automatic Sequence Controlled Calculator was spewing out volume after volume of Bessel functions of the first kind. In Aiken's group were also Lt. (later: Rear Adm.) Grace Murray Hopper (1906–1992) and Constance Franklin, mathematicians; the latter the wife and sister, respectively, of MIT mathematicians Philip Franklin and Norbert Wiener.

The system of contracts and grants in science and technology was just coming in at full force and, as I write this, is still in place. The Marshall Plan for the rehabilitation of Europe was announced by Secretary of State General George Marshall at the Harvard Commencement of June 1947, at which ceremony I received my master's degree. The transistor which totally revolutionized computer hardware was announced at the Bell Telephone Laboratories around then. The Cold War turned the USSR from a wartime ally to a bitter enemy, and the Korean War and McCarthyism surfaced in 1950.

Ed Block and the Founding of the Society for Industrial and Applied Mathematics (SIAM)

Tall, slender, and constantly wearing a dark blue trench coat of U.S. Navy issue, Lt. I. Edward Block was in the fall of 1946 one of the ten or so graduate students who, along with myself, had enrolled in Lowell Schoenfeld's course in the theory of numbers. We became friends immediately. We had some courses together and not infrequently got together socially.

Several years later in our graduate career, I recall visiting Ed one day in his dormitory room and hearing him complain that in its meetings and publications, the American Mathematical Society gave short shrift to questions of applied mathematics. I said to him that if he was so hot under the collar about this point, why didn't he do something about it. I suggested that he speak to Prof. J. L. Walsh, who was his thesis advisor and who was then either president (1949–50) or past president of the Society. Ed spoke to Walsh, who was personally quite sympathetic to applied mathematics, and I believe that Walsh may have spoken to his board about it, but nothing came of it. My recollection is that as a result of that cold shoulder, the Society for Industrial and Applied Mathematics (SIAM) was conceived in Ed's mind.

To explain the negativism that Ed (and many others) encountered, let me back up a bit. As opposed to the interests of mathematicians in, say, the U.K. or Germany, those of most American mathematicians were quite pure. When World War II came to the United States, many with mathematical training, whether of student age or older, served on active duty. Many others, however, served by working on scientific developments in universities, the armed forces, and in private industry. The work performed included developments in radar, underwater devices, atomic weaponry, ballistics, aeronautics, fluids, anti-air-craft weaponry, operations research, and cryptography. During this period, also,

Ed Block.

a number of university departments of applied mathematics were created; my own Division of Applied Mathematics at Brown dates from this period.

With the return of blessed peace, most mathematicians who had been recruited from abstract disciplines returned to those disciplines with a vengeance. They had had their belly full of mathematics in the service of creating human misery (or from sparing us from misery, as could be pointed out truthfully) and sought an escape from this potentiality by withdrawing into the world of the abstract imagination. The hostility between the pure and the applied, which predated World War II by at least a century, was thus exacerbated.

I find it sad to report that even today hostility between pure and applied mathematicians remains at a fairly high pitch, to the point that one group may "sabotage" the other group's projects. Apart from escape into the clouds of "pure reason," there are many explanations for this. They include a genuinely different view, often a reflection of personality, of what mathematics is and ought to be. The brute scramble for academic "slots" and the largest piece of a limited financial pie also play a role.

Ed's Ph.D. degree was awarded in June 1952. He moved back to his native Philadelphia and took a job at Philco. On the side, he promoted the idea of SIAM. Sympathetic scientists came forward, a little money was found, and SIAM was born. Things do get done in this world, and when you look into the matter closely, you will find someone who cares and takes responsibility. Ed Block cared deeply. A half century later SIAM is alive and well: many thousands of international members, a dozen national conferences each year, ten online technical journals, a publication list of hundreds of specialized monographs. One can find out all about the Society on its website: www.siam.org

Writing a Thesis under Ralph Boas, Jr.

The time was rapidly approaching when I had to stop taking graduate courses, write a doctoral dissertation, and then move out. These are three critical moments in one's career, each deserving of being celebrated with an appropriate rite of passage.

I followed the normal procedure, which was to find a broad area of mathematics that interested me and then find a member of the faculty who specialized in that area and who would take me on as a thesis student. My faculty advisor would suggest a topic, and then I was supposed to go ahead and dig into it on my own with only occasional consultations with him. The work would then be original work, hopefully publishable with a polite nod or acknowledgement to my advisor in a dedication or in a footnote.

In researching my thesis, in addition to the experience, the knowledge, and the advice of my advisor, in addition to my own experience and brains, I would have available the reprints my advisor gave me (usually of his own work), and access to the full mathematical collection of Widener Library. These included not only books, but mathematical and scientific journals going back to the early 1800s, and two reviewing journals: the *Mathematical Reviews*, going back to 1940; and the German language *Fortschritte der Mathematischen Wissenschaften*, going back to the mid-1800s. Before starting a project, one should always find out what is already known.

Today, the procedure for thesis writing in mathematics is not that much different. However, in addition to all of the above, the amount of computerized database information is tremendous. It is not always helpful, because it can overwhelm both the beginner and the seasoned researcher with so much information that mental paralysis results from the overdose.

In my day as a graduate student and probably going back into the deep past, there was a horror of producing a piece of work that had already been produced. We used to hear heart-chilling stories of Ph.D. candidates who completed their thesis, had it bound, and at the final moment had their hopes for a degree dashed to the ground when someone pointed out that the candidate's results were already known, and could be found, say, in the October 1926 issue of the *Rendiconti di Palermo*, or, even worse, in Vol.15 of the *Acta Societatis Scientiarum Fennicae*, 1885.

In point of fact, there is no way, even today, with all our computerization and data bases, to determine with absolute accuracy whether a piece of work has already been accomplished. Many ascriptions of original authorship are just plain wrong, even for the greatest, most original brains of the past. At a recent celebratory dinner for a mathematician at Cambridge University, a man who is a member of the Royal Society, when complimented on the amount of work he was able to do, responded publicly that it was due to his never reading the current literature. He added that to do so would have taken up so much time that he would not have been able to do his own work. Despite all this, originality remains an absolute necessity for a doctoral thesis even though this condition is not absolutely testable.

The area I wanted to work in was the theory of functions of a complex variable. Within this field (which was already vast in 1947), I wanted to work in the subfield of interpolatory function theory (rather less vast). One can see here how my high school interests and discoveries persisted and joined with my computational experiences at the NACA.

I suggested Professor David Vernon Widder as my advisor and he agreed to take me on. When I was an undergraduate, Widder was associated with Dunster House where I lived, and I would occasionally have lunch with him. I had also had a number of courses both undergraduate and graduate with him and found his material and approach congenial.

I remember Widder as a tall, athletic man, always dressed impeccably, as handsome as they come. A bit buttoned up, as one might say today, but by no means unfriendly. It was in his living room at home, at a party for graduate students, that I first heard Tom Lehrer sing some of the songs that made him famous within a few years.

Widder suggested that I look into a topic called "Lidstone series" and develop its theory a bit. He had already written several papers on the topic, which he gave me. I did not find the topic to my liking. Nor did I find Widder's suggestions as to how the theory might be deepened helpful. I floundered around and floundered around. I came up with vast sheets of formal computations that I knew would not suffice for a thesis. What was required was estimates, convergence theorems, uniqueness theorems, tie-ins with "distinguished" classes of analytic functions, tie-ins to "distinguished" transforms. These I was not able to find, (they were subsequently found by later authors), and I was desperate.

At this point Ralph Boas Jr. (1912–1992) entered my mathematical life like a *deus ex machina*. Ralph became my thesis advisor. Here is how it came about. When I returned to graduate school at the conclusion of World War II, I used to attend the Harvard-MIT mathematical colloquia fairly regularly. At that time Ralph was managing editor of the *Mathematical Reviews*, with editorial offices on the Brown campus in Providence, but he lived in Cambridge. Ralph also attended this colloquium fairly regularly and afterwards we used to walk home together up Massachusetts Avenue. From these walks and talks there ensued a friendship. I confessed to him that I was working with Widder on Lidstone series and was getting nowhere. Ralph immediately suggested another thesis topic within the theory of infinite interpolation for the class of so-called entire functions of exponential type.

The department of mathematics at Harvard was quite willing for me to work with Ralph even though he was not on the faculty. Ralph had himself been a Ph.D. student of David Widder. After he moved on, he had sat at the feet of the famous G. H. Hardy and of Norbert Wiener. As a young graduate student, I was thrilled to be apprenticed to someone so young, so brilliant, so well connected, mathematically speaking. All of us like to trace our mathematical descent from the Great Names in the business.

We met in Ralph's living room. He set a problem and gave me some references and suggestions. The material "clicked." I modified his problem, and in the course of the modification I solved a half-dozen additional problems that I had set for myself. Ralph accepted what I had done. I doubt if I we met more than four or five times. His suggestions led to my thesis in 1950 and three published papers. At the time, the theory of "entire functions" was Ralph's great love, and in 1954 he published a book on the topic that has remained a definitive text. It contains a citation to my work.

I recall Ralph as a neatly bow-tied and quite spiffily dressed fellow. With his glasses and moustache he reminded me fondly of Groucho Marx. He was slight and agile; he jumped up on desks with a balletic flair; and in the days when he was a member of the Otto Neugebauer group at Brown (Neugebauer had founded *The Mathematical Reviews*), he was nicknamed "The Squirrel"

Ralph Boas, Jr.

by Neugebauer. Many remember Ralph walking jauntily down Massachusetts Avenue in Cambridge, green bookbag (a symbol in those days of a Harvard connection) over his shoulder, on the way to Harvard Square, South Station, and then Providence through sun, rain, and snow.

Ralph had been—to use a term that was not around when he was my advisor—a "university brat." That is, his father and mother were both professors at Wheaton College, a dozen miles north of Providence. When real estate in New England was dirt cheap (in the early Thirties), and Cape Cod was more remote than it is now, Ralph's parents had bought a summer home in Orleans, which he inherited. In later years, during several summers when we rented a cottage in Truro for a family vacation with our children, we would meet together on Nauset Beach. Though Archimedes is depicted drawing diagrams in the sand, I don't recall our having discussed mathematics on the dunes at Nauset.

Modesty Is Not a Virtue:
Norbert Wiener

One day Ed Block told me that he was taking a course at MIT with Norbert Wiener (1894–1964), and why didn't I drive down with him to MIT (he had a car) and sit in on the course. I agreed; I had finished my thesis and had some time to spare. In my imagination, Norbert Wiener was a mythic figure.

I was nine or ten when I first heard of Norbert Wiener. My older brother, who was then working on an Sc.D. in chemical engineering at MIT, was taking one of Wiener's courses. At the time, Wiener was in his late thirties. I suspect now that my brother hardly needed the mathematics for his experimental thesis on the kinetics of the combustion of carbon, and that he took the course merely to say that he'd been a student of an acknowledged master.

Naturally, my brother couldn't resist telling the family about the genius at whose feet he sat. I asked him what the course was about, and I remember his response as though it had been yesterday: "Wiener talks about what the answers to certain problems would be like, assuming that the problems had answers."

Naturally, I didn't understand a word of this; it sounded like hilarious double-talk. Later, I recognized it as my introduction to the language of what mathematicians call "existence theorems."

Much later, when I listened to Wiener's course, his reputation for genius and mild eccentricity had grown; it was worldwide. He was one of those rare child prodigies who make good; he was a national figure who was soon to become a guru, an oracle, a much-quoted writer, a self-advertiser, a person whose pronouncements on such issues as the future of labor under the impact of automatization or the strategic aspects of the cold war would be well publicized. In short, in my undergraduate and graduate days, Wiener bestrode the narrow world of mathematics like a colossus and was one of only three or four colossi who did so.

Norbert Wiener (Courtesy of the MIT Museum).

The class that Ed Block invited me to attend turned out to be quite small, perhaps six or seven students. Wiener lectured, cigar in hand, and without notes. From time to time he would lay the cigar butt in the chalk tray of the blackboard. Being extremely nearsighted, (he wore thick glasses), he would often pick up a piece of chalk thinking it was the cigar. Occasionally, he would sit down in one of the students' desks in the front row, turn around in the seat to face the class, and talk a foot away from the student who happened to sit nearest him.

What did Wiener talk about? Well, the name of the course was something like "Generalized Harmonic Analysis" (a theory he'd developed in the early 1930s) but he talked about whatever he was thinking about at the moment. Feedback mechanisms and their utility in constructing prosthetic devices for handicapped people, newspapers and their sins, (he'd been a reporter briefly on a Boston paper in the early Twenties), comic strips, novelists and their sins (he'd written a novel), computers and society (this was in the first generation of electronic digital computers).

I was then working on a problem that related to non-harmonic, non-absolutely convergent trigonometric series. I got stuck on a point. Here, sitting in front of me was the world authority on the topic. One day, after class, I got up enough courage to ask Wiener about it. His answer was forthright. "I don't know the answer. It's a hard problem and I'm not aware that anyone knows the answer." I was disappointed, of course, but it bolstered my ego to know that I at least had the ability to dream up hard problems. And I learned this: never fudge an answer to a student's question.

The question I raised, call it Q1, I wanted to answer so that I could solve my "real" question, call it Q2. Ultimately I found the answer to Q2 by a totally different, much less rocky route. Thereafter, I lost all interest in Q1.

I doubt that either Ed Block or I learned much mathematics from Wiener, but what impressed me was the wide variety of things that were on his mind. In his generation, this man was among the top ten mathematicians in the world. Perhaps top five.

In the late 1940s, a scientist had to be dead between the ears to escape the talk about the social responsibility of science. The A-bomb and the H-bomb seemed to make life on earth conjectural. "Physicists have now known sin," said Prof. J. Robert Oppenheimer of A-bomb fame in an often-quoted admission of group culpability.

The cold war with the Soviet Union set in and a gush of government money to science and technology was one result. The Korean War restimulated the activities of the nation's draft boards. The House Unamerican Activities Committee (HUAC), led by Senator Joe McCarthy, pursued and harassed artists and thinkers who displayed the least softness towards socialistic ideas. Oppenheimer and numerous other scientists and mathematicians came to grief. A few mathematicians, upped, quit the country, went to Canada, the UK or Scandinavia, or to the newly founded Israel. Some never came back.

This was also the decade when the electronic digital computer blossomed and when absolutely unpredictable mathematical and technological possibilities became reality. The computer brought with it both blessings and curses, and, in its wake, generated a great deal of intellectual angst that has not yet subsided.

Norbert Wiener was in the forefront of those who raised their voices. He did it most publicly in two books, *The Human Use of Human Beings* and *God and Golem, Inc.*, and in the much less known and posthumously published *Invention: The Care and Feeding of Ideas*. It's all in these three books: the dehumanization of humans via the computer, the difficulty, if not the impossibility, of turning off thousands of types of automation once they're in place, "megabuck science," large laboratories and the rise of a generation of scientific and medical practitioners more interested in the dollar than in scientific truths.

Toward the end of his life—some have said throughout his life—Wiener seems to have been assailed by doubts about the value of his contributions to science. I have this from a number of independent sources. My own view is that he was concerned not with the short term—say one hundred years—but the long term—say five hundred years.

For the short term, his reputation was certainly secure. Even as Norbert Wiener is now fading from living memories as an individual and as a personality, his name has passed into the lifeblood of mathematics, and this process began even during his lifetime. The working mathematician now has within his grasp Wiener measure, Wiener processes, Wiener-Hopf equations, Paley-Wiener theorems, the Wiener extrapolation of linear time series, and Wiener's generalization of harmonic analysis.

Wiener was the man who put the word "cybernetics" into our current vocabulary, and every child who watches TV or gets on a computer terminal knows who the "cyborgs" are and what kind of "cyberpunks" live in "cyberspace." If you are over 18 (ha, ha) you can experience "cybersex" on the Web. The prefix "cy" now represents something that is simultaneously desirable, fearful, futuristic, supertechnological. Wiener himself would very likely have been revulsed by the contents of our current cyberworld, let alone the projected cyberfuture.

Opinions of the Famous:
Buttonholing Buber

Tv, whose technological principles were known quite early, came in slowly, commercially speaking; it didn't really get off the ground in the United States until after World War II. I recall walking up Massachusetts Avenue on my way home after class and seeing small clusters of people watching a TV set turned on in the window of an electrical shop. The people on the sidewalk were watching a boxing match or a baseball game. A neighbor, somewhat older than I, told me that he was going to start a TV magazine. I said to myself, he'll be out of business in two months. I was blind. There's a foundation now that goes by this man's name.

More striking was this experience: sometime during the years 1948–1952, my father-in-law wrote us that if we showed up for lunch on a certain Saturday afternoon, we would have the pleasure of meeting Martin Buber. Buber (1878–1965), a short, white-bearded Santa Claus type of figure, was then at the height of his world fame as a Chassidic storyteller, an existentialist philosopher, and a theologian. In my mind, he was to the world a super-guru. His *I and Thou* (*Ich und Du*, 1923) dialogue theology advanced the idea that a dialogue between God and man is possible despite the infinite distance between them, and this belief exerted considerable influence on Christian theologians.

Of course my wife and I could not forego this experience, and we turned up for lunch. Buber and his wife were both there, along with a few other older people. I have no recollection of what the general conversation was about, only that I remained fairly mute during it all. Lunch over, people got up and started milling about in the living room and I found a chance to buttonhole Buber.

He asked me what my profession was and I told him. He replied that he had always liked mathematics and mentioned a number of famous mathematicians in Germany that he'd known. Then he asked me why mathematics is important today, and I said, just off the top of my head, "Well, take television, for example.

Martin Buber (© Gütersloher Verlagshaus, Gütersloh).

The mathematical basis of television transmission is tremendous." Then I asked him whether he had watched any TV; he said that he had. I asked him what future he foresaw for it, and he answered that he thought it would be a tremendous boon to education. He elaborated on this for a bit, and shortly thereafter another person came over and a different conversational tack was pursued.

A great boon to education! Yes. Maybe. Possibly. What has TV become in the intervening half-century? The greatest boon to education as well as to miseducation, to information as well as to misinformation, the greatest threat to social relationships. It has also been a technological and behavioral ally of the personal computer, the Internet, the Web, which, again, are touted as media through which an absolute revolution in education will come about. This may very likely be the case. The meaning of education may change rapidly, and if we were able to fast-forward to a hundred years from now, we would probably not even identify what we see as education.

My father-in-law used to tell me that I was a bit of a mystic. In response, I confessed to him a bit of antirationalism but I said it didn't add up to much. In having lunch with Buber, I had hoped to hear about mysticism "straight from the horse's mouth," so to speak. That didn't work out, but I didn't abandon the hope of a direct word from someone who had mysticism in his blood (but not a person of the Madame Blavatsky school).

A few years later I was able to spend an hour with a younger man who was a charismatic figure in his own right: Abraham J. Heschel (1907–1972). Heschel received his doctorate from the University of Berlin in 1935, having been let through (miraculously) by the rabid Nazi dean and mathematician Ludwig Bieberbach. When he arrived in the United States, Heschel's multifold activities extended from writing the book *God in Search of Man* (1955), to marching, garlanded in a lei, shoulder to shoulder with Martin Luther King in the front line of the famous Selma March ten years later.

Fundamentalism, said Heschel, claims that the "big questions" have all been answered. Positivism claims that the "big questions" are meaningless. (Philipp Frank would have denied this.) Heschel seemed to propose something halfway, by replacing a philosophy of doubt with a philosophy of mystic wonder. I wanted to hear about it.

Had I ever experienced wonder as I studied and learned and created new mathematics on my own? I certainly felt early on the wonder of the strange phenomena that mathematics elicited; e.g., that every whole number is the sum of at most four square numbers, or that there are precisely five regular convex three-dimensional polyhedra.

However, as I studied these statements more deeply, particularly in their historic context, and learned their proofs, I began to feel rather as the great Flemish mathematician and engineer Simon Stevin van Brugge (1548–1620) did when he provided an almost visual, kinesthetic proof of the law of the inclined plane. In a famous illustration in his 1605 *Wisconstige Gedachtnissen*, he surrounded his triangular inclined plane with a closed chain, which obviously, he said, would not rotate. If you remove the symmetrical part below, the upper portion of the chain will still be in equilibrium, and the weight of the two portions of the chain will be in proportion to the lengths of their respective sides of the triangle.

Below this figure, Stevin added the caption (in Flemish) "*WONDER EN ES GHEEN WONDER.*" In English: A wonder, and yet it's no wonder at all.

Yes, I've felt great wonder in the statements of mathematics, and yet, I have experienced the waning and the disappearance of that wonder as I've succeeded in understanding what lay behind the statements. Rationalism and wonder are in conflict.

Despite my father-in-law's judgment, I don't take to mysticism of any sort, whether Eastern, Western, Existential, neo-Pythagorean, pagan, or New Age, very easily, and all I brought home from my encounter with Heschel was a "homework assignment" that I have yet to do. Heschel told me: "Read Plotinus, you will enjoy him."

Cut the Cackle, Professor

I assume that some social anthropologist has already devised a scale that measures the degree of snobbism in a particular profession. On my own personal scale, I would give my mathematical contemporaries an eight or nine on a scale of ten. The university at which they earned their degree, the prominence of their degree supervisor, the current cachet of the subfield and sub-subspeciality within the subfield, the university at which they are now employed (God forbid that they should take a job in secondary education, industry, government, or set up their own business or that they should abandon mathematics entirely!), the depth of their research work (zero depth if they "merely" teach), all are vital parameters that determine who should snoot whom. What Sigmund Freud termed "the narcissism of small differences" was the order of the day in my mathematical generation.

When I declared mathematics as my undergraduate major, I never in my life thought that I could make a living at it. I majored in mathematics because I liked it, because I was good at it. The Hot War that began on September 2, 1939, the Cold War that began practically the moment the Hot War was over, and the Technological War that began when Sputnik was launched on October 4, 1957, the tremendous computer revolution that was *in utero* even as I was computing at the NACA with the aid of a slide rule: all these changed the employment possibilities for mathematicians completely. But academic snobbisms still held sway.

However, it was not all a smooth ride as regards getting a job. I naturally inherited a bit of the snobbism described above. I would give myself a three in that regard; lower than most because during World War II I was doing aerodynamics at the NACA at Langley Field, Virginia.

When I received my Ph.D. degree in 1950, I wanted a university position. At the moment my diploma was handed me, very few university jobs were available, and those that were, were not exactly desirable even from the point of view of my reasonably low snob scale.

I recall going for a job interview at a certain northeastern university I shall call Springlock. The mathematics department at Springlock University was a "very good one," which meant it placed great emphasis on faculty research. This was a plus. I was interviewed by the chairman of the department, whom I shall call Professor Lindwurm. I could see, even on entry, that Professor Lindwurm was impressed by my credentials, and that he would like me to join his staff.

He showed me around the premises; a very nice campus indeed. The football stadium was not too large: a plus. The mathematics department was housed in an old Victorian residence with ample office space for all. Another plus. Then he began talking about himself; how he lived seven or eight miles out of town and raised chickens and corn. I agreed with him that alimentary independence was a noble goal and he was pleased to hear my concurrence. He kept on talking about his chickens and his corn and how some of his corn was of the right sort to feed to his chickens.

He said nothing whatsoever about the salary I could expect. The message came through to me slowly but sharply and I didn't have to be Sigmund Freud to interpret it: my salary would be chickenfeed and I would soon resemble Hogarth's *Distressed Poet*, who sacrificed his wife, his child, his cat, and her kittens to the seductions of his Muse. I went back home and wrote Professor Lindwurm saying thank you, but no thank you, and Lindwurm very likely concluded that I was a wiseguy kid.

A second interview at a Midatlantic University I shall call Springbok. I gave the colloquium presentation required of all job candidates and passed with flying colors. Prof. Wurmlind, the chairman of the department, then took me out for the standard post-colloquium meal. He apologized that we would not be joined by other members of the department. (A bad sign.) He asked me whether I wanted to eat Chinese or Italian. I chose Italian. "An excellent choice," he said. (A good sign. I could have the job if I wanted it.)

We drove in Wurmlind's car to a place called Luigi's. I ordered spaghetti. Wurmlind waxed poetic about how Luigi's mother made the spaghetti at home and hung it up to dry on her clothesline. "Homemade spaghetti is the only kind worth eating," I ventured. "Agreed," replied Wurmlind. (A good sign.) We finished up. Still no mention of salary or of the number of courses I would be required to teach. (A bad sign.)

After some months of additional scrounging around and taking a number of short, unsatisfactory jobs, I found through Stefan Bergman's advocacy, a job at the National Bureau of Standards in Washington, a traitorous move away from academia, and one that I'm sure raised eyebrows in the academic circles in which I had been traveling.

Part V

Mathematics in Washington

What Is the Case and
How Do We Know It?

The early Fifties were not wonderful times for hiring mathematicians, and I was grateful that I was able to secure a position at the National Bureau of Standards in Washington. (NBS; now NIST: National Institute of Standards and Technology.) Even then, it was not entirely a straightforward engagement, for I was hired not by the NBS but by American University which, in turn, was under contract to the NBS. A year or so later, my anomalous position was changed to regular civil service employment. And even then, over the next decade, the employment situation at NBS was iffy. From time to time there would be layoffs of its scientific staff.

When I signed on, the NBS had been "under a cloud" for four years. The Bureau was set up by Congress in the late 1890s as a laboratory in the service of technology and general scientific research for the standardization of weights and measures. The prototype yard and meter bars were on public view, displayed in a large safe on the first floor of the Administration Building.

The technological exigencies of two world wars, particularly the second, led to a great expansion of the NBS's activities in scientific research and development of many sorts. With the close of World War II, the NBS did some product testing—probably as an offshoot of their mission of standardization. I think that if the general public had ever heard of the NBS, they thought of it as the place where the government tested certain consumer goods for quality; in point of fact, I don't know how much of this was actually done. In the late 1940s, product testing at the NBS got it into trouble in the famous "Battery-Additive Case."

Stated briefly: a certain business man put on the market some kind of mixture called AD-X2, which he claimed extended the life of car batteries.

The claim was challenged and the NBS was asked to investigate it. The director of the NBS at the time was the distinguished physicist Dr. Edward U. Condon
(1945–1951), and the head of the statistical section of the NBS was Dr. Churchill
Eisenhart, a man of world reputation in statistical theory and practice.

The NBS tested the product. The substance AD-X2 was found to consist
largely of some common chemical. The life of a battery right off the production line is itself probabilistic. It was found that, on average, the battery additive was ineffective. The manufacturer then claimed that the presence of some
"trace elements" did the trick and put pressure on the NBS by working through
congressmen. The upshot: Condon resigned or was forced out. The NBS got
out of product testing, but I believe it never retracted its findings in the AD-
X2 case.

When I arrived in September, 1952, the shooting was over but the echoes
of the battery additive were still reverberating. When I would tell a non-
scientist that I had taken a job at the NBS, like as not they would say "Oh, the
battery-additive place." The scientists at NBS were slightly depressed to learn
that scientific inquiry was subject to commercial and political pressures. After
all, the USA was not supposed to be like the Soviet Union, where biological
research had been set back greatly by Stalin's favoring Lysenko's theories.

It was at this point in my life that the difficult question entered my mind of
the degree of subjectivity or objectivity that resides in scientific experiments
and theories. The question, generally, is: what is the case and how do we know
it? In such a general context, the question is hardly answerable, and I have
voted steadily for a mixture of objectivity and subjectivity: the absolutism
inherent in claims of one hundred percent objectivity has always repelled me.

This question was hardly raised in the lectures of Philipp Frank I attended
in college. It was certainly raised by the positivist philosopher Alfred Ayer, as
will be reported later in this book.

Though the question has interested me mainly in the context of mathematics, I have always been dissatisfied with explications of the so-called
scientific method. Observations, formulation of theories, predictions, experiments, verification of the predictions? That's the way science is usually
presented in elementary texts. A little more subtle and satisfactory is the
"falsification theory" of philosopher Karl Popper, that a statement cannot be
scientific unless it is potentially falsifiable by experimentation. But even
that view has holes in it.

Around this time (mid-Fifties) I once remarked to David Park, a friend
who is both a quantum physicist and a historian of physics, more or less as a

joke, that after much study and cogitation I had arrived at a satisfactory definition of the "scientific method."

"What is it?" he asked me.

"Try everything you can think of," I answered. David did not have violent objections to my formulation, but added a clause.

"Yes, try everything, but don't try anything stupid."

Condon took a job at the University of Colorado. In 1966 he headed up an Air Force commission to investigate UFOs! In January 1969, it released the famous (and controversial) "Condon Study" containing negative findings.

> "Our general conclusion," the report said, "is that nothing has come from the study of UFOs in the past twenty-one years that has added to scientific knowledge....Further extensive studies of UFOs probably cannot be justified in their expectation that science will be advanced thereby."

Condon was once again into hot water—this time with the "paranormal community," among others.

I think that the fraternity of philosophers may be bored with the general question. Yet it does not go away. What is the status of psychoanalysis? Of acupuncture? Of alternative medicines, of homeopathic substances, of a universal fluid called the ether, of cold fusion, of proximity to high tension wires and cancer, of witchcraft?

I am reminded of a letter received some years ago from Professor Donald Avery from a site in Africa where primitive methods of metallurgy were still being practiced. Avery noted that part of the procedure for producing high quality iron involved a number of virgins marching up and down in a fixed and complex ritual.

How can one tell what are the necessary parts of a technology? What is good science and what is bad science; what is real science and what is pseudoscience? What is sincerely performed and what is a conscious and deliberate hoax? Many answers, many criteria have been proposed; none is necessary, none is sufficient.

Power Galore: Talking with Leland

When we moved from Cambridge to Washington in 1952, a real estate agent found us a house in the area known as Chevy Chase. It turned out that the house belonged to Leland Olds, a name I knew from the newspapers. A few years previously, as the then chairman of the Federal Power Commission, Olds had been subjected to a Senate inquiry.

The house was a large one, quite adequate to the needs of a family with two children, and my wife and I were well satisfied with it. It was sufficiently close to the National Bureau of Standards, so that every once in a while I would walk to work.

It was a common feeling among us residents of Washington in the Fifties and Sixties, that Washington was the center of the universe. The city was full of elected lawmakers and their staffs, appointees, civil servants (such as myself) beyond count, lobbyists, the military, the diplomatic corps and their families, writers, analysts, think tankers, journalists, wealthy hostesses. The atmosphere was both heady and jittery. We used to say that local news was international news. And so it was.

There is a famous definition of a diplomat given by Sir Henry Wotton (1568–1639): "An ambassador is an honest man sent to lie abroad for the good of his country." A less cynical definition was given to me in the mid-Fifties by a man who, with his wife, was on the fringes of the vast social buzz in Washington: "A diplomat is a person who is able to say graciously to his hostess 'What a lovely cake.'"

During and after World War II, the city grew enormously. We would meet natives who recalled that in their childhood, people took summer cottages near where we lived on Nebraska Avenue (about three-quarters of the way to the northern apex of the District). In our day, cows were pastured outside the Walter Johnson High School in Bethesda.

The people I dealt with professionally and the people my wife and I met socially were all—whether Americans or not—highly educated, and I came to think, rightly or wrongly, that Washington existed primarily as an absorber of the overproduction of such people by our native colleges and foreign universities. Are universal education and the growth of bureaucracy inevitably coupled?

In those pre-Selma Days, the large black community of the city was pretty much ignored by the white community. More than that, the white community seemed to be strangely and paradoxically isolated from the rest of the United States. When I questioned a colleague who was a born-and-bred Washingtonian, he told me that the city had not suffered at all during the Great Depression of the '30s. I concluded that Washington took care of its own (except for the black community) first and foremost and the rest of the country was lower down on the totem pole.

We had a year's lease on our house. Toward the end of the year, the agent told us that Mr. Olds was moving back to Washington, and would not renew our lease. Without much difficulty, we found another house a block or two away, and during the switchover, we got to know Mr. Olds, his wife Maude, and his two daughters Zara Olds Chapin and Mary Olds. In my eleven-year hitch in Washington, Olds was one of the closest I got to the inner (appointed) circles of the government.

I found Leland Olds a tremendous conversationalist; on his part, he was interested in the work that my colleagues and I were doing at NBS (this was during the first generation of electronic computers), perhaps, in part, because his father, George D. Olds, had been a mathematician at Amherst College (and later, president of Amherst).

From time to time, my wife and I would walk up the street to our old house and talk to Leland. He struck me as simultaneously a technocrat and a socialist of the native American variety; a man in the tradition of Walt Whitman, Henry George, Eugene Debs, Thorstein Veblen, and a man who was doubly secure in his views because of his genealogy. He told us that after he got married to Maude, he built a house for them to live in, and that the rude mission-style dining room table, whose use we enjoyed while we were renting, was an example of his craftsmanship.

Prudence and thrift were built into Leland's character. Zara told us that once, when he and his wife were invited to a White House function, he drove up in his battered old Ford, and then handed it over for valet parking. When I knew him, Leland wore cowboy boots and a ten-gallon hat, and one might think that he was a native of Oklahoma (which Maude was), rather than a New Englander.

He was in favor of public power: the TVA, that sort of thing. He dreamed of a series of publicly owned hydroelectric plants on the Missouri River. The dream came true and one of the large Western power plants in Stanton, North Dakota, is now named in his memory.

Leland expressed both endless delight and complete exasperation when he told us about the dirty tricks the automobile industry indulged in in the 1920s to compete against the substantial American railroad and street railway systems. During this time, railroad managements, fat, complacent, and resting on their ancient congress-granted privileges, milked their assets instead of ploughing money back into renovation. Immediately after World War I, very few could have foreseen America's and the World's love affair with the automobile, just as immediately after World War II, Tom Watson, head of IBM, could not foresee the later accomplishments of computers and the world's love affair with them.

Leland Olds (1890–1960), served on the Federal Power Commission from 1939 to 1949 and most of this time he was its chairman. Given his outspoken nature and his opposition to the country's power barons, he accumulated enemies in the Senate. In 1949, President Truman nominated him for a third five-year term, and when his nomination came up for Senate approval, the investigators pulled out of the archives material he had written years before, which criticized the free enterprise system as it then operated. The Senate turned down his reappointment overwhelmingly. It is rumored that after his inquisition at the hands of the Senate Committee, his neighbors would hardly talk to him and Maude.

At the time I knew Olds, magneto-hydrodynamics was a hot topic among both technologists and theoreticians, and it was well-supported financially because it seemed to offer the promise of cheap power beyond all limits. I was not engaged in any MHD work, as were some of my graduate school classmates who specialized in partial differential equations and as were some of my Bureau colleagues. I knew just enough to keep Leland informed on what was happening in the field.

Together we speculated on what unlimited quantities of absolutely cheap and non-polluting power would mean to the world. Leland was inclined, but not absolutely, to the rosy view: that mankind would benefit; that the standard of living would rise; that nations would be equalized; that power struggles between nations would be mitigated. On the other hand, my inborn skepticism suggested to me that unlimited power would be an unlimited disaster. Populations would skyrocket, national social and economic structures in which a finely balanced equilibrium had been achieved over centuries would collapse, controls would be ineffective, and general chaos and destruction would prevail.

Luckily—so far—neither of our views has been put to the test, for magneto-hydrodynamics, both for technological and theoretical reasons is currently a dead duck. Nor, despite the cold fusion flap and flop at the University of Utah, has technology come up with an alternate idea for cheap, non-polluting power in unlimited quantities. However, the dream was alive and well for a good decade after Leland and I used to talk about its social consequences.

As an instance, and here I look forward a bit. In 1965, shortly after I joined its faculty, Brown University dedicated its newly built Barus-Holley Physics and Engineering Building. At the inauguration of the building many notables were invited. The key address was given by Sir Fred Hoyle, the world renowned Plumian Professor of Astronomy at Cambridge University, cosmologist and an advocate of something called "panspermia" (the theory that life on earth was seeded from outer space). In his address, Hoyle latched on to the idea of unlimited power available to mankind and in a sci-fi dream that he revealed to his audience, he utilized this power, but not to provide us all with goodies such as villas in Southampton and unlimited trips to Cancun. Sir Fred suggested that the power be used to move the orbits of the planets into a more felicitous configuration! For what purpose he wanted to make this mid-cosmic correction, I now have no recollection.

Top o' the Morning to You, Prof. Synge

As a kid I loved to spin tops. I mean the kind of top you wind up with a long string and throw to the ground while retaining the string in your fist. Tops were "in"; you could buy a top for a nickel. The string cost a penny. All my friends spun tops. When I was in top form (so to speak), I could hold two tops in my hand and spin them simultaneously.

As this skill began to diminish, I began to collect all kinds of tops; little tops that came free in a box of Cracker Jack, wooden whip tops, tops that worked on a spring, big metal tops that you spun by pushing down on a spiralled shaft, toy gyroscopes, yo-yos, diabolos; any toy that spun rapidly. My collection has been dispersed but its hum lingers on.

When I joined the Bureau of Standards, there were many distinguished visitors to the mathematics division who stayed for longer or shorter periods. There was also a shortage of room space. For a while I had to share an office with several such visitors and the very first was Professor John Lighton Synge, FRS, (1897–1995), out of whose book I had studied theoretical mechanics. (Synge and Griffith: *Principles of Mechanics*, 1942)

Synge was a companionable sort, an easy roommate. We had our stretches of silence during which we worked, and we had stretches when we chatted. I heard much about his older relative John Millington Synge (1871–1909), the well-known Irish playwright. I had read J. M.'s *Playboy of the Western World*, so I was able to energize J. L.'s narratives by asking a few intelligent questions.

I was with Synge for about a month and then he went on to other engagements. During this time, Hadassah had brought home a little top of a sort that I had never seen, intending it for our children. The top was about an inch and a quarter in length and made of plastic. Its body was a hollow sphere, or more precisely, a sphere with its "north polar cap" missing. Built into the sphere and emerging from it was a long, narrow cylindrical shaft by which the top was spun between forefinger and thumb.

But this was by no means an ordinary top, for its spherical shape—as opposed to the classical conical shape—enabled it to do a great trick. When spun, the top would continue to spin for a while in the normal fashion. Then, very gradually, the cylindrical shaft would approach the ground, and then, miraculously, so it seemed to me, its spherical body would rise, and the top would spin upside down on its shaft and even "go to sleep" in that position. How to explain that!

Well, as I have said, I was sharing an office with one of the world authorities on mechanics, and I recalled that in Synge and Griffith there was a long chapter on the motion of a rigid body in space. Vector analysis, cross products, gyroscopic couples, that sort of thing. I would bring the top in, spin it on Synge's desk, and ask him if he could account for the strange and wonderful behavior of my top.

In the morning, I did just that. Synge observed the motion of the top carefully. Then, rather non-commitally, he said something like "Very nice, very nice," and went back to the manuscript of the book he was then working on: *The Hypercircle Method in Mathematical Physics* (1957).

The next morning I found on my desk a folio-sized reprint of an article in the *Transactions of the Royal Irish Academy*, bearing an inscription to me. Author: J.L. Synge. An approximate title: "On the Motion of the Tippe Top."

Well, topped by the Master! I thanked him and promised to read his article through. I did this over the next few days and was able to have my "revenge."

"If I'm not mistaken, Prof. Synge, you've added a frictional term to the usual equations."

"Yes, a non-linear frictional term."

"And that explains why the top doesn't go to sleep in the normal position. The normal position is unstable."

"Correct. That explains it."

"But your paper doesn't explain why the top turns over and goes to sleep upside down?"

"You're right. I'm afraid it doesn't. That would be a very difficult problem. A global problem. Prediction in the short term is often easy. But in the long term...aahh...."

Yes, aahh... I let it go at that. But now I will ask: what has happened to the theory of the tippe top since the time I shared a room with Synge? In the intervening years the theory of dynamical systems has grown tremendously. Tops are again popular as an example of a "completely integrable system." Can the complete motion of the tippe top be derived theoretically from the principles of dynamical systems?

I asked my colleague Prof. Constantine Dafermos. His reaction was, "The tippe top is probably constructed with a high center of gravity and this would make the lower position less stable than the upper one. I doubt whether the complete motion can be derived. But I will see what's around."

Working on Abramowitz-Stegun

During the tail end of World War II, in my job as an aerodynamicist at the NACA, I worked with all kinds of mathematical tables. I remember in particular, using the recently published table of exponentials computed by the WPA Tables Project in New York, a project that was set up in the late Thirties as an aid to unemployed mathematicians. In those pre-digital computer days, when the word "computer" meant a person and not a piece of hardware, several thousand special functions had already been computed and published. One of the prestigious compilations in those days was Jahnke-Emde's: *Funktionentafeln mit Formeln und Kurven* (1909, and many later editions).

In 1954, a committee, chaired by Professor Philip Morse of MIT, recommended the planning and publication of what was to become *The Handbook of Mathematical Functions* (1964, U.S. Department of Commerce, Applied Mathematics Series 55; often called "AMS 55"). I was drawn into the effort immediately by Milton Abramowitz, who was in charge of the project, and I volunteered to prepare a chapter on the gamma function. Why the gamma function? Because as an undergraduate reading period assignment in complex variable theory with David V. Widder I selected as my topic the complex gamma function. So why not take advantage of my knowledge?

Actually, I contributed two chapters to the book: the "Gamma Function," Chapter 6, and "Numerical Interpolation, Differentiation and Integration," Chapter 25. As the first author to get my chapters in to the editors, I believe I forged a certain format that other contributing authors found useful in laying out their own work.

The big, fat, red handbook appeared in 1964 (skillfully guided to completion by Irene Stegun after Abramowitz's premature death) and was an immediate success. It sold more than 100,000 copies in the Government Printing

Office version and probably twice or three times that number in legally "pirated" versions. It has been my most successful publishing project if one judges by sheer numbers.

Money? The writers (such as myself) who were on the staff of the National Bureau of Standards, made their salaries. Those who were not had a contractual arrangement. No one made one penny of royalties either from the government edition or from the legally "pirated" commercial edition that appeared later. We were government employees at the time the work was performed and to accept royalties for work performed on governmental time was considered double-dipping.

When Congress passed a special law allowing President Dwight D. Eisenhower's various writings to gain him royalties (*The White House Years* 1963–1965), we were all irritated. In my ungenerous view, Eisenhower's initials D.D. at that time stood for double-dip.

But money does not totally rule the world. I recall visiting the Department of Applied Mathematics and Theoretical Physics in Cambridge, England a few years ago. One morning, I had an hour to kill, and I walked into the mathematics library. All was quiet, people were working with books and notes. From a distance I could see that one reader (much older than the students) had AMS 55 open in front of him. My pride and curiosity got the better of me and I sidled over to where he was sitting and peeked over his shoulder. He was poring over my pages on the gamma function. Can one measure in dollars the value of the occasional thrill that comes from authorship?

Though some people might have regarded table making as mathematical Grub-Street hackwork, I did not. Apart from my desire to deepen and then to display my knowledge of the gamma function, there was another reason for my readiness to join the *Handbook* project. We are all born into a world that we didn't make. As students and as researchers, we go into a mathematical world that is already there, that was created by hundreds and thousands of people stretching back into the mists of time. This world nurtures us professionally, and from it we may forge a career. The world then expects something in return. To some, the only possible return consists of a bundle of technical papers. Others see the necessity of increasing the comprehensibility, the availability, the utility of what the previous person has done. My feeling is strong that one ought to return something utilitarian to one's discipline even if it has a rote quality.

I believe this feeling came from my working my doctoral thesis under Ralph Boas. His early association with the *Mathematical Reviews* which I witnessed, and his subsequent career that embraced writing advanced monographs, text books, editorial work, and Russian language mathematical dictionaries, argued strongly for the merit and the seriousness of doing community work within the world mathematical community.

The making of mathematical tables and formularies of special functions is as old as recorded history. The so-called Rhind Papyrus contains a mathematical table that goes back almost 4,000 years. Its purpose, expressed in modern terms, is to facilitate the expression of fractions as sums of fractions whose numerators are all 1.

Reader, did you know, or do you care to know that $2/13 = 1/8 + 1/52 + 1/104$? Check it out if you want to. For reasons best known to themselves, the ancient Egyptian mathematicians adopted this inconvenient way of representing numbers and made extensive tables of such decompositions.

A "special" mathematical function is one that is useful in many different mathematical and applied mathematical contexts. To name a few: the power functions (squares, cubes, square roots, etc.), the logarithmic function, the trigonometric functions, the exponential function, the exponential integral, the Bessel functions, etc. One of these special functions has even entered into the common language; one often reads that such and such has experienced (or will experience) exponential growth, meaning, in the technical sense, growth that is characteristic of compound interest, but in the popular sense, simply very rapid and solid growth.

Don't I remember vividly the multiplication table (up to 12 x 12) that was printed on the back of my copybook? Didn't I, as a high school boy, practice for hours with tables of logarithms and trigonometric functions? Didn't I, with an adjoining table of proportional parts, laboriously interpolate to values that were not listed in the tables?

There are now thousands of special mathematical functions whose properties have been studied and are still being studied. Over the years, the finest mathematicians, Ptolemy, Kepler, Legendre, Gauss, Karl Pearson, to name a few that come to mind easily, have devoted a part of their career to creating tables. How to access and use "special" functions profitably is an art in itself.

In the years since its publication, the *Handbook* has been referred to in thousands upon thousands of articles. (It is often called "Abramowitz-Stegun.") And yet we authors were aware, even as we read and wrote and selected and compiled and arranged and computed, that computerization was moving so rapidly that the book would contain many obsolete features of content, of format, and of usage even before it hit the stands.

Tables now have been replaced by computer algorithms and formal identities are often incorporated within high-level commercial computational languages such as MATLAB, MAPLE, and MATHEMATICA. Books themselves are being replaced by Web sites.

It has now been proposed to "multimedia-tize" AMS 55 and this job has been taken on by the mathematicians at NIST, the National Institute for Science and Technology, the successor of the old Bureau of Standards. There are many ideas floating around as to how best to do it. These ideas compete in a mathematical marketplace that combines both computer technology and strong commercial motives. My only advice to the current generation of table makers is to make sure that their intellectual property will be protected. For better and for worse, mathematics has now become a commercial product.

The Adams-Jefferson Letters

Like any resident of Washington, D.C., or any visitor to the city, my wife and I were put in contact, one way or another, with Thomas Jefferson (1743-1826), the third President of the United States. Driving through the Tidal Basin, we saw from our car the magnificent Jefferson Memorial, dedicated during the presidency of Franklin D. Roosevelt. Visiting one of the government buildings on Constitution Avenue, we lined up with hundreds of tourists and took a quick peek at the Declaration of Independence, a document that derived, for the most part, from Jefferson's pen.

Stimulated, perhaps, by these and other evidences of the past that inheres in the city, Hadassah became interested in early American history. She was later able to carry her interest forward to a higher academic degree and to writing some books on the subject.

In 1959, the North Carolina University Press brought out a two-volume collection, edited by Lester J. Cappon, of the letters of John and Abigail Adams to and from Thomas Jefferson. John Adams was, of course, the second President of the United States. Hadassah saw an ad for the books, ordered them, and read them through.

As far as I was concerned, the books were Hadassah's property and they lay around the house unopened by me. I thought that all they contained were dull collections of official governmental documents or the angry recriminations of two men whose political visions and careers were often in conflict. Several months later, out of simple curiosity, I flipped through them. I read and was amazed with what I found. I kept on reading. Since that day, I've kept Cappon's two volumes near my bed.

The letters in these volumes include an extensive correspondence between Adams and Jefferson long after both men had left active politics. Jefferson

John Adams (courtesy of the John Carter Brown Library, Brown University).

was in retirement at Monticello and Adams was in retirement at his home in Quincy, Massachusetts. They had put aside their political animosities. These letters explore the world of ideas: society, government, history, literature, science. The correspondence is well worth reading today for its insightful perceptions of two men of action and two thinkers who have left distinct marks on history. Adams, the older of the two men, writes more informally, loosely; he seems to initiate more of the discussions. Jefferson replies more formally.

In February, 1819, Adams, who had little inclination to theoretical science, sent Jefferson a collection of papers of Nathaniel Bowditch (1773–1838), an American mathematician, navigator and astronomer, and raised the question of what they were all about. After a few weeks, Jefferson responded as follows:

> I am in debt to you for Bowditch's mathematical papers, the calculations of which are not for every reader, altho' their results are readily enough understood. One of these impairs the confidence I had reposed in LaPlace's demonstration, that the eccentricities of the planets of our system could oscillate only within narrow limits, and therefore could authorize no inference that the system must, by its own laws, come to an end.

The work mentioned in Jefferson's response, together with that of Nathaniel Bowditch, is the *Mécanique Céleste*, the masterwork of the Marquis

Pierre-Simon de Laplace (1749–1827), a man who with considerable justification was known as "the Newton of France." I was thrilled when I read this response. Possibly no American scientist could read these lines of Jefferson without experiencing a thrill that one of the founding fathers of the United States was a man of such training and disposition of mind that he could talk about the ideas of celestial mechanics with some understanding.

I was particularly engaged because of my own slight experience with celestial mechanics. It came about in this way. In the late Forties, a candidate for the degree of Ph.D. in mathematics at Harvard was required to write a major and a minor thesis. Whereas the major thesis had to be a piece of original work, pushing forward the mathematical sciences in some direction, the minor thesis was written on an assigned topic and had to be completed within three weeks. It could be a survey or an exposition of work already done in that topic.

My minor thesis topic was assigned one afternoon when I walked into the office of Garrett Birkhoff and told him I felt ready to take on the job. He reached into his desk and drew out a little slip of paper the size of the fortune that comes in a Chinese fortune cookie. Perhaps he had lots of such slips and drew one out at random. I opened the slip and read: "Celestial Mechanics." That was all. Two words. I thanked him and walked out of his office.

The two words overwhelmed me. I experienced hot flashes and cold chills. I felt I was swimming and then drowning in an endless sea. Celestial mechanics was one of the oldest topics in the history of astronomy and mathematics. Hipparchus, Aristarchus, Apollonius and Claudius Ptolemaeus had written on it in ancient days; much later, Copernicus, Tycho, Kepler, Galileo, Newton, Euler, Lagrange, Laplace. Great, great names all. In the nineteenth century, Bowditch, Hamilton, Hill, Weierstrass, Poincaré; and in the early 20th, George David Birkhoff, a mathematician of world class and the father of the man who handed me my "fortune" topic. Still later, the famous Russian mathematicians Kolmogoroff and Arnold, and the German mathematician Moser.

I selected an unoccupied carrel in the bowels of Widener library, poorly lighted and far from the madding crowd, and began my reading. I resisted—it was not difficult—wasting time reading the books in the adjacent stacks, which were on the classics in the Pali language. When I realized how much material on celestial mechanics there was, hundreds and hundreds of papers and books, I felt absolutely lost. Those two words, celestial mechanics, were impossible for me or for anyone else to cover in three weeks or three months or even in a lifetime.

Recovered from my initial shock, I realized I had to cut the topic down to human size by selecting one small theory within celestial mechanics and describing it as best I could. Should I describe ancient material? I discarded that possibility; I was well aware that the mathematical establishment despises its own history except when a question of priority arises.

I would have to select a topic that was fairly recent and one to which someone of repute had given high grades. G.D. Birkhoff's work? I had once flipped through his 1927 book *Dynamical Systems*. Well, I really didn't want to summarize Birkhoff's work. I figured that would be too close for comfort. But reading through the introduction to Birkhoff's book, I found a sentence that told me what to do: describe Karl Sundman's (1912) theory of planetary collisions. Here is how I put the matter in the introduction to my minor thesis, dated June, 1948:

> The theory of Sundman is a particularly good one for treatment in this paper of some fifty odd pages. In support of this I suggest the following reasons. In the first place, it is a modern theory and one which, in the words of G.D. Birkhoff, is "one of the most remarkable contributions to the problem of three bodies, which has ever been made." Yet it has received less attention than the topological methods inaugurated by Poincaré. Furthermore, the development is to a great extent self-contained. Lastly, the subject is of great intrinsic interest. Indeed, in these days of cosmic sensationalism, what could be more *à propos* than a theory of planetary collisions?

Well, my nod to the elder Birkhoff worked. I went to Sundman's original papers, *Mémoire et Récherches sur le Problème des Trois Corps*. My exposition of his theory was accepted. I proceeded without delay to my major thesis topic— miles removed from celestial mechanics. By the time of Laplace, celestial mechanics was applied mathematics; and mathematics, to the general populace, was one of the least understood, least appreciated, most misconstrued of the various intellectual disciplines. The general ignorance of this beautiful subject—a tree of ideas with ancient roots and modern fruit—is profound, and is beset with fear, odium, superstition, misinformation, and cosmic sensationalism.

As I write, recurring allusions in the newspapers to collisions with meteorites and comets, combined with religious fanaticisms and apocalyptic millennialisms, show that cosmic sensationalism is around "in spades." It rakes in billions for the movie makers. As for myself, a black hole is not an object to be combined with dinosaurs and cyborgs who are in possession of meta-tech-

Thomas Jefferson (courtesy of the John Carter Brown Library, Brown University).

nology and have the morals of pre-cavemen. It is largely a mathematical construct that has physical implications.

But back to Jefferson. Jefferson's reply to Adams made me think: Just how much mathematics did Jefferson know? How much astronomy? How much science? Most historians have little interest in science, and so this aspect of Jefferson is not generally explored.

A full description of Jefferson's preoccupation with science and technology would require a full volume, and the interested reader will profit from looking into the books of Silvio Bedini, Edwin T. Martin, and I. Bernard Cohen. It will be seen that there was hardly a science or a technology, natural or physical, about which Jefferson did not inform himself. From comets to vaccination, from the growing of white wines to the determination of longitudes, from the ancient history of the Indians to the construction of geometrical models, all was grist to the mill of his restless curiosity and imagination.

In the scope of his learning, wide interests and personal accomplishments, Thomas Jefferson was undoubtedly the most intellectual President who has yet sat in the White House (1801–1809). Throughout his life, though not a scientist in the professional sense of the word, he immersed himself in science and technology; though not a mathematician, he knew rather more mathematics than the average educated person of his day. As President, he fostered the support of science by his young government (e.g., the Lewis and Clark Expedition), a course of action that was by no means popular with the politicians and in some ways contradicted his own advocacy of minimal government. In his day, Jefferson could very well have served as his own science adviser. And, of course, he did.

My curiosity then spread from the man to the response he gave Adams: how, if at all, had that topic developed since 1819? The answer in a few words: the topic is alive. It's well. It's called "the stability of dynamical systems." It's one of the specialties of my own Division of Applied Mathematics at Brown University. It is related to computing, to chaos, to the limits of what can and what cannot be predicted by mathematical means. And if a dynamical systems theorist wants to travel the road of cosmic sensationalism, deriving The Beginning of it All and The End of it All from the equations, and deriving God en route, it's very easy to do so.

Thomas Jefferson:
An Applied Mathematician in the White House?

As part of my increased interest in Jefferson, in the spring of 1960, my family and I visited Monticello, Jefferson's home atop an eight-hundred-foot mountain outside of Charlottesville, Virginia. The ride from Washington is an easy one, and the location in beautiful Charlottesville, in the heart of the Blue Ridge Mountains, offers more to do for a family with small children than merely looking at architecture and thinking about ancient(!) history. The Skyline Drive and Luray Caverns are not far away.

We stayed in an old and famous southern hotel in downtown Charlottesville, its gracious dining room staffed with black waiters in livery, its biscuits kept warm in a special silver-plated heater. The traces of the old regime before the civil rights movement were still visible.

I had seen a few other presidential residences: those of George Washington (Mount Vernon), Franklin D. Roosevelt (Hyde Park, New York), John Adams (Quincy, Massachusetts) and John Quincy Adams (also in Quincy). It struck me at once that of these, Monticello was by far the most expressive of the personality of its occupant. In Monticello, one is immersed immediately in the presence of its architect and first resident. For starters, there is the building itself, designed by Jefferson in the Palladian style, but exhibiting a geometric grace that is unique.

The building is palatial, but not on the European scale. Next to Sans Souci or Versailles, Schoenbrunn or Blenheim, it is a modest dwelling. Monticello was the master's residence on a working plantation, palatial in the sense of its contrast to what could then be found in rude and rural America not far from the frontier of the vast wilderness to the west.

In the interior, one finds a variety of special features, Jeffersonian mechanical inventions and gadgets, that suggest the interests, the hobbies, the

Phil Davis on the steps of Monticello, 1956.

conveniences of the inventor. I recall seeing a trick bed, a dumbwaiter in the dining room, a clock and windvane, a pantographic device for making multiple copies of documents. The various memorabilia, the eyeglasses, the music stand, add to the sense of the personal, and the local guides, with their constant references to "Mr. Jefferson this and Mr. Jefferson that," completed the illusion. It was easy to believe that the retired president might at any moment issue from one of the smaller rooms in his robe and pantouffles, greet the tourists as his guests and offer them a glass of wine from his private imported stock.

As we left Monticello, we passed through the inevitable gift shop, selling copies of the *Jefferson Bible* in which Jefferson anthologized the Gospels, deleting what he considered superstitious or miraculous, and retaining what he considered its ethical core. Also on sale was Claire Burke's sweet-smelling "Potpourri" of dried flower petals.

Outside the house at Monticello, one sees the far off hills; the city of Charlottesville is hardly visible, and with a little imagination one can be back 250 years to the open country and to the great importance of the mathematics of surveying.

Jefferson's father, Peter, who, among many other things, surveyed and mapped a vast unexplored portion of western Virginia, liked mathematics and introduced his son to it. In college at William and Mary, in Williamsburg,

Virginia, Thomas Jefferson met up with William Small, a faculty member who influenced him greatly. Small, a Scotsman with an unorthodox mind, was his mathematics teacher.

The mathematical college curriculum at that time consisted typically of the elements of algebra, geometry, surveying and navigation. William Small opened up science and mathematics for Jefferson in a most wonderful way. Among the mathematical texts that Small used were George Adams's *The Construction and Use of Terrestrial and Celestial Globes* (1719), the *Philosophiae Naturalis Principia Mathematica* of Sir Isaac Newton (1687), and William Whiston's *Sir Isaac Newton's Mathematical Philosophy More Easily Demonstrated*, (Forty lectures read in the public schools of Cambridge. London: 1716).

Jefferson, like his father before him, worked for a while as a surveyor. He was a skillful architect. He made maps, he collected mathematical books and mathematical instruments—which, in the classification of his day, ranged from drawing instruments through clocks, thermometers, and astronomical devices. In France, he met among other scientists and mathematicians, Lagrange and Condorcet. He was impressed by the *Mécanique Analytique* of Lagrange which pulled together, formalized and generalized Newton's work on mechanics. Lagrange took Isaac Newton's work out of the geometrical domain and into the algebraic or analytical domain (i.e., calculus) and applied it (among many other things) to the motion of fluids.

Jefferson loved to study mathematics and he said that it had been the "passion of my life." In a letter to Pierre du Pont written in 1809 just as he was leaving the Presidency, he explained why mathematics had given way to stronger passions:

> Nature intended me for the tranquil pursuits of science, by rendering them my supreme delight, but the enormities of the times in which I had lived, have forced me to take part in resisting them, and to commit myself on the boisterous ocean of political passions.

In this respect, many young people with a mathematical bent, before and after Jefferson's day, have found themselves in a similar personal dilemma. Talented, brilliant even in many fields, but realizing that the pursuit of a particular passion required an exclusive and exhausting expenditure of mental energies, they are forced to make a decision. "What road shall I follow?" was a question dreamed by young René Descartes on the night of his scientific revelation, and the answer he gave changed the world forever. Jefferson, who might very well have taken up mathematics, chose a different career, a career that also changed the world, but he kept a permanent watch on mathematics out of the corner of his eye.

In a letter to a P. K. Rogers written in 1819, Jefferson was sufficiently knowledgeable to recognize the brightest and the best in the mathematical world: Euler, Bézout, Lacroix, Legendre, Monge, Laplace; and to give reasons, in terms of the inner development of the calculus, why England had lagged behind France in the pursuit of mathematics.

Much of Jefferson's retirement energy went into the founding of the University of Virginia. He designed a number of its buildings and was responsible for giving mathematics a prominent place in its curriculum.

Two direct mathematical accomplishments of Jefferson are his plough and his cipher cylinder. In the first instance, he was concerned with devising a more efficient plough than was commonly available, and to this end devised a mathematical surface consisting of wedge-like shapes in series. His method allowed him to generate a surface "whose characteristics will be a combination of the principle of the wedge in cross directions, & will give what we seek: *the mould board of least resistance*."

An applied mathematician today would not accept Jefferson's claim that he had devised the plough of least resistance. The problem would be reformulated in terms of fluid theory, the calculus of variations, and this would be combined with a bit of computer-aided geometrical design. Despite the fact that efficient ploughs have been designed and are commercially available, any theoretician attempting the problem today in a way that would be consistent with the standards of theory would find that adequate underlying physical principles are still missing.

To quote a letter from Trevor Stuart of Imperial College, London, and a Member of the Royal Society on this point:

> There are papers on optimization (minimum drag) in flow problems. E.g., O. Pironneau, J. Fluid Mech., 64, 1974, p.97 and R. Glowinski and O. Pironneau 72, 1975, p. 385. The second is a numerical paper. These papers are for very slow speeds, (linearized, i.e., Stokes' equations—a biharmonic, steady problem). I do not know of much for high speeds, (high Reynold's number) for airplanes, birds, fish, etc. Pironneau's

work is relevant to small organisms. As I recall, the optimal axisymmetric body of minimum drag, has, under some constraints, a pointed fore.

Soils are difficult. There is really no satisfactory theory of soil or sand movement. But R. A. Bagnold, (Army Brigadier and brother of the playwright Enid Bagnold) did lots of nice work on the physics of desert sand. It might be possible to do something on ploughs by a fluid analogy. G. I. Taylor did nice work on the mechanics of painting and paint brushes. Fishes may be relevant!

Jefferson's duties (1789–1792) as Secretary of State in Washington's presidency, and the necessity for privacy of communication in negotiations led him to cryptography. The Jefferson-Madison Correspondence in the 1780s, while Jefferson was in France, was coded. He devised a cryptographic scheme and built a cylinder with rotatable alphabetic segments to facilitate the use of the scheme. In point of fact, Jefferson's scheme was identical to one that had been devised in the late sixteenth century by the French cryptographer de Vigenère. It comes under the heading of polyalphabetic ciphers and today it is not considered secure if a large body of enciphered material is available for analysis.

Cryptography is now a highly mathematized subject, drawing on deep results of the theory of numbers, and is of great practical importance given the necessity for privacy in the computer transfer and processing of information. To quote from a paper of Smith and Yeh,

> Protection of information is one of the most important issues facing computer users today. This is because of the increasing number of applications involving network communications and distributed systems, the growing amount of information processed by computers, the increasing dependence on shared databases, and the increasing number of applications critical for the existence and success of enterprises. Cryptography is an effective tool providing data security (preventing unauthorized access to data), data integrity (ensuring the accuracy of data, and authentication (verifying the identity of a user or sender).

Cryptography has now become a major topic of mathematical research of which, for reasons of national security, only a part of its theory and practice is in the open literature. An abiding question is how secure a given cryptographic system is, and this depends in great measure on how rapidly a computer can compute.

One of the most highly regarded cryptosystems is known as RSA (Rivest, Shamir, Adleman, 1977) and is employed in numerous commercial computer applications. The mathematical basis of RSA, Euler's theorem in the theory of numbers, was around in Jefferson's day. Once the broad features of a system are known, the cryptographic community goes to work and tries to devise "attacks" on the system which would destroy its security. As I write, the following judgment on RSA was made by Dan Boneh (*Notices of the American Mathematical Society*, February, 1999):

> Two decades of research into inverting [i.e., decoding] the RSA function produced some insightful attacks, but no devastating attack has ever been found. The attacks discovered so far mainly illustrate the pitfalls to be avoided when implementing RSA. At the moment it appears that proper implementation can be trusted to provide security in the digital world.

With "quantum computers" now being developed in theoretical papers if not yet realized in hardware, current cryptographic systems would rapidly become liable to successful decoding attacks, and "quantum-cryptographic" schemes would be called in to provide adequate security.

James Joseph Sylvester

On the modest obelisk that marks the grave of Thomas Jefferson, one can make out in weatherbeaten letters the three things for which he wanted to be remembered: as author of the Declaration of American Independence, and of the Statute of Virginia for religious freedom, and as the father of the University of Virginia.

In the early 1820s, the "Sage of Monticello," now an old man, sat on the veranda of his spacious mansion and peered through his telescope at construction going up several miles away across the hills. The new University of Virginia in Charlottesville was being built. The university was the most important achievement of Jefferson's post-Presidential life. He drew up the architectural plans for its central buildings just as years before he had designed his own house, and when we visited its campus, we immediately felt the presence of the President.

Though residing in remote Charlottesville, Jefferson stood out as Sirius among fourth or fifth magnitude stars. The world knocked at his door with letters of introduction; he put up guests by the dozens and wined them and dined them. Lawyer, revolutionary, democrat, governor of Virginia, diplomat, author, twice President of the United States, a patron of the sciences and a slightly more than amateur scientist, an eighteenth century philosopher, a planter, an agricultural experimentalist and a manufacturer of nails, this extraordinary man dreamed of a university that would have a worldwide reputation.

Jefferson was also a Virginian among Virginians, and I have the strong impression that his neighbors in Charlottesville didn't much care for him. Genius is envied, his taste for French wines seemed precious, his religion was minimal: one of ethics and probably scandalous in a region that would subsequently be known for its fundamentalism. Although he was a slaveholder himself, his democratic views were outrageous in a society that was based on slaveholding. In short, he was a traitor to his class, just as a century later President Franklin D. Roosevelt was held to be.

Jefferson knew what first-rate minds were and he wanted them for his university. Why shouldn't he have them? Hadn't the great natural scientist Alexander von Humboldt visited Monticello but a few years back? He wrote letters and offered appointments. But the scientists stayed away. Here is how in 1825 he described the situation to John Adams:

> I am glad to learn that Mr. Tickner has safely returned to his friends. But should have been much gladder had he accepted the Professorship in our University, which we have offered him in form. Mr. Bowditch too refuses us. So fascinating is the *vinculum of the dulce natale solum* [the bond of sweet natal soil]. Our wish is to procure natives where they can be found, like these gentlemen, of the first order of acquirement in their respective lines; but, preferring foreigners of the 1st. order to natives of the 2nd. we shall certainly have to go, for several of our Professors, to countries more advanced in science than we.

> I am and shall always be affectionately and respectfully your's.

> Th: Jefferson

Adams was of a different opinion:

> I do believe there are sufficient scholars in America to fill your Professorships and Tutorships with more active ingenuity, and independent minds, than you can bring from Europe. The Europeans are all deeply tainted with prejudices both Ecclesiastical and Temporal which they can never get rid of.

In the short run, Adams's view prevailed. What established European scholar would abandon the comforts, the resources, and the intellectual stimulation at home for the uncertainties and the rigors of an institution that had been newly established at the edge of the wilderness? What promise of growth and stability would the University of Virginia have after the passing years had inevitably withdrawn the loving care of its founder? Charlottesville was then too remote even for native American academics hailing from Boston or Philadelphia.

Jefferson died in 1826, and after Madison, Monroe, and other Virginians of the old stripe passed away, the plantation mentality took over. The code of honor prevailed and hot bloods that sought trouble and demanded personal satisfaction populated Jefferson's university.

Listen to what Henry Adams, the brilliant New England Brahmin, historian, crank, and part-time bigot had to say about several Virginians who were his classmates in the Harvard Class of 1858:

> Strictly the Southerner had no mind; he had temperament. He was not a scholar; he had no intellectual training; he could not analyze an idea; and could not even conceive of admitting two; but in life one could get along very well without ideas, if one had only the social instinct... when a Virginian had brooded a few days over an imaginary grief and substantial whiskey, none of his Northern friends could be sure that he might not be waiting round the corner with a knife or pistol, to revenge insult by the dry light of delirium tremens.

Into a student body that must have been composed in some measure of such bloods, there walked in 1841, fifteen years after Jefferson's death, a most unusual professor. He came from England: St. John's College, Cambridge. He was 27, he was unmarried, (he never would marry), he was talented (he was Second Wrangler and became one of the leading mathematicians of his generation), and he was a Jew. As such he was excluded from an academic degree and preferment because of his inability to sign the Thirty-Nine Articles of Faith of the Church of England. His degree came from the University of Dublin. What did he have to lose by coming to the young country?

I think the ghost of Jefferson was probably well satisfied with James J. Sylvester's appointment to his university. But the truth of the matter is that Sylvester didn't last long in Charlottesville. Chalk it up, if you will, to a conflict of like temperaments. Sylvester was a short, stocky man, of tremendous vitality and of a fiery passion. The denouement, or more strictly, the end of Chapter I of Sylvester in America, came four years later.

In one of his classes Sylvester had as students two brothers who must have been of the precise type described by Henry Adams. One day, Professor Sylvester criticized the recitation of the younger brother in such a flowery statement of deprecation that the boy's sense of honor was offended. He sent word round that Sylvester must apologize or suffer the consequences. Sylvester did not apologize. Instead he bought himself a sword cane. The student got himself a heavy walking stick. The two brothers set an ambush for the professor, and when Sylvester walked into it, the younger brother demanded an apology, then, almost immediately, knocked off Sylvester's hat and hit him on the head with his heavy stick. Sylvester drew his sword cane and pierced his assailant over the heart.

The boy fell back into his brother's arms and cried out "I am killed." A spectator came up and urged Sylvester away. He left immediately for New York without waiting to pack his books. The boy was not hurt seriously. The point of the sword had fortunately struck a rib.

If Sylvester could have stuck it out in Charlottesville, then *caeteris pari-bus* as they say, Jefferson's university would have been the alma mater of important developments in the theory of matrices (with which I have spent more than a little time), of invariants and hosts of other mathematical conceptions. But Charlottesville would have to wait a bit longer for its hour in the sun.

Sylvester tried to get jobs at Harvard and Columbia. They turned him down. As it worked out, the United States had to wait until 1876. In that year, after a career in the Royal Military Academy at Woolwich, Sylvester accepted a chair in the newly founded Johns Hopkins University in Baltimore. There, shortly thereafter, Sylvester began the *American Journal of Mathematics*. With that, American mathematics came of age.

Around that time also, the importation of foreign mathematicians of the first class, mentioned in Jefferson's letter to Adams, became much much easier and rather less necessary.

Emilie Haynsworth

Jefferson's South could not in his day have fostered a woman mathematician—nor, indeed, could almost any place in the world. The number of women mathematicians up to Jefferson's time who have left any substantial record behind is few. Mention Hypatia of Alexandria (d. 415), Gabrielle Emilie Le Tonellier, Marquise du Chatelet (1706–1749), Maria Agnesi (1718–1799), Sophie Germain (1776–1831), Mary Somerville (1780–1872), Ada, Countess of Lovelace, Byron's daughter, (1815–1852) a few others perhaps and you have pretty much exhausted the list.

After 1850, women mathematicians, while rare in proportion to men, become more common. After World War I, women appear thick and fast on the mathematical scene. There are now websites that contain databases of such names. Although women suffer from discrimination and are underrepresented, there are a sufficent number of them to be able to form "separatist" groups. Some feminist mathematicians have come to consider the current corpus of mathematics as a macho production and speak of a possible femininization of mathematics.

I once read that the famous playwright Lillian Hellman, who was born in Louisiana, would have foregone all her dramatic successes to have been a southern belle. Not the Emilie I knew. Emilie Virginia Haynsworth (1916–1985), a most feminine lady, abandoned her status as a southern belle from Sumter, South Carolina, and concentrated on what in her generation was still considered rather unfeminine—mathematics. Directly, she was responsible for intensifying my interest in matrix theory, and indirectly in the relation between mathematics and society.

One of the very first things I learned about Emilie, she having rapped my knuckles for it, was that she was an "e-less" Haynsworth.

"Well," I countered, "I'm a one *l* Philip. Less is more, you know."

"It surely is." (She pronounced it "showly.") That settled the matter and sealed a friendship of some thirty years.

I first met Emilie in Washington, D.C. When I joined the National Bureau of Standards in the early 1950s, I was part of a group that was rethinking and reworking numerical methods with full attention paid to the potentialities of what is now called the first generation of electronic digital computers. Besides the permanent staff there was a constant stream of distinguished mathematicians coming through. Many were from Europe, and at one time there were so many from Switzerland that the joke went around that we ought to raise the Swiss flag over the Bureau's administration building.

Several years after I arrived, Emilie joined the group. I worked on several projects with her, and together we wrote a joint paper on the condition number of matrices.

In those years, Professor Alexander Ostrowski of the University of Basel, a mathematician of great international reputation, was an occasional visitor at the Bureau of Standards. Ostrowski had turned his mathematical attention from complex variables and valuation theory to matrix theory, Emilie's specialty. He and Emilie struck up a personal and professional friendship that lasted the rest of her life. Through him she got to know many continental mathematicians and spent many productive months in Basel and Prague. She had a fine sense of mathematical elegance—a quality not easily defined. Her research can be found in a number of books on advanced matrix theory under the topic: the "Schur Complement." Emilie taught me many things about matrix theory.

For a short while, we shared an office, and we talked about everything from matrix theory to politics to how one of her collateral Haynsworth ancestors, as a young man, fired the first shot at Fort Sumter in the Civil War against the Yankees. On both sides of her family, Emilie came from a long line of South Carolinians. She was a member of the first families of South Carolina with all the grace and the charm that are classically associated with the South. Mathematics was what fired her imagination and she made her mark in matrix theory.

She knew some Gullah, a dialect of remote black South Carolinian islanders. She'd learned it as a child from native speakers, and every once in a while she would wrap up a situation succinctly by producing a proverb in this strange and colorful language. Since I got interested in the language, she gave my wife and me a cookbook, produced by the Junior League of Charleston, South Carolina, in which each chapter is headed by an epigraph in Gullah.

Emilie Haynsworth.

Emilie liked Washington. She liked the North. She had spent a bit of time in Maine, and for a while was a graduate student at Columbia University in New York City. Ultimately, the pull of university teaching and the lure of her native soil led her to take a job at Auburn University in Auburn, Alabama (where, she told me later, Miz Lillian, President Jimmy Carter's mother, was a popular house mother in one of the frats).

"So you're returning to your native soil," I said to her when I heard the news.

"Nonsense. That's just your northern provincialism. Alabama is *not* my native soil."

And then she gave me a lecture in history, in geography, and how the banks of the Pocotaligo and Chattahoochee Rivers each introduce discontinuities into local pronunciations. She then moved on to genealogy, concentrating, of course, on the "*e*-less" Haynsworths. "There are certainly Haynesworths with an *e*," she informed me with a twinkle in her eye, "but of course we don't talk to them. Thank God."

Hadassah and I were sad when Emilie left Washington, but our friendship continued. In the spring of 1962, at her suggestion, I was invited to give the Sigma Xi series of lectures at Auburn. I said I would lecture on the philosophy of mathematics concentrating on mathematics and society. Emily agreed to the topic. This opportunity required that I think through a bit more carefully just what my philosophy was. These lectures were among the first I gave on the relation between mathematics and society, a line of thought from which I've not yet extricated myself.

In those years, there was still a train connection from Washington to New Orleans that stopped in Auburn. I took the train. Emilie met me at the little station which is now, I believe, a boutique. A departmental party for me was held several days after I arrived. I knew very little about Auburn. I could see that it was a small college town, but I very rapidly came to realize that the university was a tremendous powerhouse in college football. There was a gap in the general conversation, and I turned to Emilie next to whom I was sitting, and asked her, "What's special about Auburn?" (Meaning the town and not the university.)

"Nothing much. It's just a small town."

"Oh, come on. Every town has something special about it."

"Well, if you really want to know," she answered me loud and clear so the whole living room could hear, "Auburn is the only place south of Nashville that has a drugstore where you can buy a horse collar on Sundays."

I suspect that as the years went on, Emilie became a bit of a character on the Auburn scene. Living with her dogs in her modern, prefab octagonal house on Gold Hill, close to the birds, the cattle, and the thundering freight trains, she firmly resisted attempts of well-wishers to move her into town where many thought she would have been more comfortable in view of her declining health. She clung without complaint to her privacy and her independence. She drove wherever she wanted to in her Ford, and contributed much to my nervousness by always driving it around on "empty."

She was cut from the same tough bolt of cloth as her mother, who, she once told me proudly, had been the first woman to scale the Matterhorn. Her students all called her Miz Emilie in affection, but my New England ears cannot detect all the overtones that are present in this mode of address. (This was long before Miz or Ms. became a catchphrase of women's liberation).

In the spring quarter of 1984, at Emilie's suggestion, I gave a course at Auburn on The Nature of Mathematical Thought. Everywhere, and not just at Auburn, boys were not encouraged to take courses on the humanistic aspects of science; nuts-and-bolts courses are what the boys opted for. As a result, my class was full of beautifully groomed and graciously mannered southern belles, each fully capable of winning the Miss America contest, and most majoring in nursing or some very practical study. Coming to my lectures faithfully, Miz Emilie sat among the undergraduates taking notes.

One day, possibly spurred on by the particular topic I was discussing, she said she wanted to give me a present of something she treasured. She brought it in the next day. It was a letter her father had written her in the late Thirties when she was about to begin her teaching career. Her father had been a lawyer with some teaching experience as a young man, and had put together a number of precepts for his daughter to follow when she stood in front of her class. I'm sure she did.

She was a fairly devout Episcopalian. She was a political and social liberal. She was not vocal, but her beliefs and actions were firm. A remote cousin, Clement Haynsworth, a judge in South Carolina, had been nominated in 1969 by President Nixon for membership in the Supreme Court. The Senate turned him down. "I'm glad they killed his nomination," Emilie said to me quietly, "he's just a terrible conservative."

I am thinking now of one of the last times I saw Emilie. It was in her house at Gold Hill. She was hostessing one of those post-colloquium parties known locally as "War Eagles" (the Auburn University mascot). She sat, very frail, nursing a bourbon, neat, and enjoying her guests. She had as little self pity as one can imagine in a person. She loved her dogs, who nearly did her in when they tripped her up one day. To her, their action had been one of brutish love and was no reason for cursing.

Emilie spent one summer in Providence. We worked on a textbook on matrix theory, which was completed, and which I used in my lectures, but which for a variety of reasons, was never published. She lived with her nephew Pete Haynsworth and his wife Ruth in their East Greenwich, Rhode Island, house. (East Grunch, Emilie called it.)

While she was here her old friend Helen White from South Carolina flew down from Camden, Maine to visit her. Helen wanted to see a bit of what she called the odd-ball New England scene. So we went on a pilgrimage to Adamsville, R.I., where there is a monument to a chicken—the Rhode Island Red, a breed famous in the history of American poultry.

Emilie was contemptuous. "In my part of the country we'd never put up a monument to a chicken."

"What *would* you do?"

"We'd fry it!"

Chief Justice Warren and
Professor Niels Bohr

The influence of science and technology on society is clearly visible. Look at what gunpowder, the steam engine, electricity, the automobile, atomic energy, the Salk vaccine, the computer, the Internet have done to the way we live, to the way we organize our lives. Not to mention the zipper and other such trivialities.

The other way around, the influence of society and its social and ethical systems on science is much harder to delineate. Some people even deny that it can be done sensibly. One such denier—or hesitater, at any rate—was Otto Neugebauer. Whenever I would bring up the topic at lunch time, he would make a rude noise: his way of telling me that I was talking *quatsch*. (When I told this story to a historian of science in Vienna, my auditor said: "Oh, Neugebauer. What did you expect? He was from Graz!")

Why did mathematical progress stop for 1,000 years or more? Why did the decimal system appear when it did? What were the forces of society that led to Fourier series, to non-Euclidean geometry? What keeps mathematical research going? Isn't there enough of the stuff around already? Answers can be attempted but are not entirely convincing.

I doubt if the famous Danish physicist and the Chief Justice of the U.S. Supreme Court ever met except in my own imagination. One day in the mid-Fifties while I was living in Washington, my father-in-law called me up from New York and told me that he had an appointment with Chief Justice Warren late on the next afternoon. Following his appointment he would come up to Chevy Chase to visit with us a bit. Then, as an afterthought, he added, "Would you like to meet the Chief Justice?"

"Of course I would."

"In that case, then, why don't you come down to the Supreme Court Building at such and such a time. I'll be winding up my business. You'll meet him and then we'll go back to your house."

Earl Warren, former Governor of California, was named to the court by President Eisenhower in 1953, and by the time of the conversation just recorded he had established a reputation as one of the most liberal justices. The term "Warren Court" was used both as praise and as condemnation.

Parking not being easy in downtown governmental Washington, I decided to take the bus. I got off the bus and walked to the front of the Supreme Court Building. In all the years I'd lived in the city, this was the closest I'd been to it. Classical architecture, graceful proportions, a long set of steps leading to a portico, and on either side of the steps, there was something I had never noticed in photos of the building: two large ornamental basins filled with water. I couldn't tell whether the basins were catch basins for fountains.

These two basins together with my knowledge that Warren had been recently subject to much adverse comment put me in mind of a verse I knew from the Old Testament: "Let loving kindness flow as a river and justice as a stream." (Amos, 5).

My personal interpretation of this parallelism is that it points to the unresolved tension or the conflict that exists between the formal law (justice, judgement) and the human, informal qualities of mercy and tenderness (loving kindness). In a flash, as I walked up the flight of steps, I imagined that I when I met the Chief Justice, we would exchange a few words, and then it being the end of a long day, we three would walk out together, he to his limo and we to our taxi. In the course of which, we would pass the two water basins, and (clever me) I would quote the verse of Amos to him by way of praise that he had managed to balance the two conflicting ideals.

Well, it didn't work out that way. I walked in the front door of the Supreme Court Building. I was met by a uniformed usher. I gave him my name and stated my business. The usher phoned to the Chief Justice's chambers. I was expected. He led me to the chambers. My father-in-law, already with hat in hand, introduced me to Earl Warren, whom I recall as a very relaxed and informal sort of man. We exchanged a few polite words about where and what my job was. He replied with the response most often heard by mathematicians: "Well, I was never much good at mathematics." The meeting was over. My father-in-law and I were ushered out. And I never had the chance to quote the words of Amos.

Forty years have now passed in the twinkling of an eye, and in preparation for my sketch of Philipp Frank in a previous section, I reread some of Frank's writings. In a collection of talks given on the occasion of the retirement from teaching of both Percy Bridgman and Philipp Frank, (*Science and the Modern Mind* (1958) I found an article ("The Growth of Science and the Structure of Culture") by J. Robert Oppenheimer, theoretical physicist, leader of the Manhattan Project, and later director of the Institute for Advanced Study at Princeton. In a part of this article, Oppenheimer attempts to relate various scientific advances with the ambient culture. Here is how Oppenheimer dealt with Bohr's principle of complementarity in quantum mechanics:

> Bohr participated perhaps more than any one man in the development of atomic mechanics. To the decisive formal discoveries of Heisenberg, Schroedinger, Dirac, he was very close. Yet he has told me that his interest in the ideas of complementarity long antedated these discoveries in atomic physics. They sprang from his interest in the complementary character of the introspective and the behavioral description of man, in the complementary character of dealing with experience in the light of love and in the light of justice, and from the familiar yet disturbing tensions of comprehending in one description causal explanation of behavior and moral condemnation of behavior. These traits of our experience...are neither easy to ignore nor easy to resolve.

Oppenheimer associates here the scientific imagination with man as a moral being, and selects love and justice, two aspects of being that come close to my Warren story.

Communication Difficulties and Solutions

Krazy Kat: Why is lenguage, Ignatz?
  Ignatz Mouse: Language is that we may understand one another.
K.K.: Is that so?
  I.M.: Yes, that's so.
K.K.: Can you unda-stend a Finn or a Leplander or a Oshkosher?
  I.M.: No.
K.K.: Can a Finn or a Leplander or a Oshkosher unda-stend you?
  I.M.: No.
K.K.: Then, I would say, that lenguage is, that we mis-undastend
  each udda.
  - *Krazy Kat: the Comic Art of George Herriman* (1986)

If the autobiographical portions of this book have given very few specific details of what I think I've accomplished within "hard core" mathematics, the omission has been deliberate. There are at least two reasons for it. If the reader is a specialist in the areas I've worked in, he will conclude with some justification that my contributions to those areas have been pretty slight. And as regards the general reader, if I had provided details, they would not be understood.

In this section, I shall distinguish three kinds of mathematical writers and audiences: the mathematical specialists, the more general mathematical practitioners, and the educated laity. Admittedly these groups are vaguely defined because, for example, a specialist in one field may not be a specialist in another field. This is commonplace; in medicine, for example, your G.P. does not usually do orthopedic surgery. When the members of the groups communicate, they may write for themselves or for another group; and I shall talk about several intergroup communication problems.

Specialist-Specialist

When the specialists have the same speciality, there are few communications problems. They have common experiences, knowledge, insights and goals; they have a common vocabulary, notation, and manner of exposition. When the specialists have different specialties, problems develop. Many expository papers and lectures are given by specialists at large meetings that cover a number of specialties. I have found that few of them succeed in communicating anything to me. After a few nods in the direction of easy introductory material, the writer or lecturer resumes his professional specialist's stance.

Specialist-Mathematical General Practitioner

If the specialist can unbend to the point where he assumes no more than the course work of the average graduate student in mathematics, the paper or the talk will be rather more successful in reaching the larger audience. The writer or the speaker must discard any thoughts that by such a presentation he loses status as a deep and original thinker. He must abandon the idea that non-intelligibility will be interpreted as profundity. Various professional mathematical institutions offer a number of prizes for such expository papers and talks and I have won a number of them. The carrot helps but the problem abides.

Specialist-General Educated and Interested Public

Here there is a severe communications problem: how to convey to the average person (and I include high school students here) the contents of current mathematical research. If the work done can be expressed briefly in terms of simple ideas of arithmetic, there is no problem. But this limitation buys us rather little. The statement of the notorious "Fermat's Last Theorem" can be explained to everyone who knows about addition and multiplication of numbers. The proof of the theorem is beyond the working knowledge of the average professional. In principle, the professional might study up and see what is going on there. In practice, it is rarely done, except by those few Fermat "buffs" who have devoted years to it.

Computer graphics has added a significant new dimension to the arsenal of the expositor. Videos, CD-Roms, computer demonstrations both numerical and graphical can, in some cases, serve to give a feeling for what a particular problem and its solution are all about. But these new media are subject to the same problem: the "script writer" frequently assumes too much knowledge of his viewers and develops the subject too rapidly.

The communications gap has been noted for a long time. It was already an old question when in 1819 John Adams sent Jefferson a copy of Bowditch, and Jefferson, in turn, gave Adams a pretty good lay explanation. The first notice in print of the communications gap that I am familiar with was written two centuries ago by a clergyman in the German town of Woltershausen in the State of Niedersachsen.

> The great mathematicians do not allow themselves to make their science comprehensible to us beginners. Is it because they imagine that others have the same degree of understanding that they do and hence they consider it superfluous to expound those statements that form the bases of their arguments? Or is it that they consider that those who do not have as much sagacity as they do are incapable of understanding the mathematical sciences and therefore they think it not worth their while to instruct them? Or is it that they feel it is unpleasant and repulsive to explain matters which, in their opinion, are just too simple?
>
> - Georg Ludewig Spohr, *Anweisungen zur Differential- und Integral Rechnung fuer Anfaenger*. Leipzig, 1793. Reprinted in *Drawbridge Up: Mathematics—a Cultural Anathema*, H.M. Enzensberger (1998).

I'm not sure how the late Anneli Lax of the Courant Institute of Mathematics in New York City got my name, but sometime in the late Fifties she wrote and asked whether I would write a volume in a new series of mathematics books that she was editing. At the time, I knew her and her husband, the mathematician Peter Lax, only slightly. I was, of course, flattered, and asked her for more information.

She responded that there was a general reluctance on the part of research mathematicians to tell the wider public what they knew and how they viewed mathematics. She had begun a series of books called *The New Mathematical Library* whose aim was to bridge this communications gap. She had already lined up a number of authors and the first books in the series had already appeared. Would I be willing to join this effort?

I answered immediately that I was quite aware of the general reluctance, that I thought the plans for her series were admirable, and that I would write such a book. More than that: I told her that if it met with her approval, I would write on the simplest topic for young people that I could imagine. Numbers;

particularly large numbers or long numbers, thinking that such things would be especially mysterious or funny and hence attractive.

The Lore of Large Numbers appeared in 1961 as Volume Six of the *New Mathematical Library*. Over the years it has had a modest sale. It is still in print, though badly out of date; it was written before computers became commonplace. Perhaps the peak of my pleasure with the book came many years later. I was giving a colloquium lecture at a university in Ohio and was introduced to a young man who was chairman of the statistics department. He told me that as a boy he had grown up on *The Lore of Large Numbers*.

It was my first book and I could hardly have guessed (despite my penchant for writing) that over the next four decades I would be producing books, technical, expository, philosophical, fictional, at about one every three or four years. These books have won me numerous awards and have reached people in many strange places, including a high-security prisoner in Colorado, and political prisoners in Poland and Lebanon.

The New Mathematical Library has flourished, through a number of publishers, and it now lists more than forty titles. There is much less reluctance than there used to be on the part of research scientists to try to reach out to the general public. Editors and literary agents sniff out prospective authors and entice them with the prospects of large royalties. Despite the diminution of snobbism that put an onus on those scientists who "go popular," the communication problem is still with us.

Lay-Lay

Consider now the case of Hans Magnus Enzensberger. My university library has sixty of his works on its shelves, yet the name of Hans Magnus Enzensberger (b. 1929) is not widely known in the United States. In Germany, he has a great reputation as an essayist, a journalist, and as one the country's most prominent poets. His latest book is *Zig Zag: The Politics of Culture and Vice Versa*. Enzensberger's political writings are controversial and academic conferences have been based on them. Having now read in translation a number of his books, I have found him to be a fine writer and much more easily understood than most of the academics who have sought to explicate him in conferences.

At the 1998 International Congress of Mathematicians in Berlin, Germany, Enzensberger gave a Urania talk, i.e., one held in the Urania Theater which is dedicated to "high culture" at a popular level. The talk was immediately published in the *Frankfurter Allgemeine Zeitung*, perhaps the leading high-brow newspaper in Germany. It appeared in the USA in an English translation in the book just mentioned that contains the quotation from Spohr.

If Enzensberger is an "outsider to mathematics," as he has himself stated, what are his qualifications for writing about mathematics? I asked him this question and he wrote back to me

> My mathematical training is negligible. I have been following, to the extent that they are accessible to a layman, the developments in mathematics, as a sort of intellectual fitness training. I had the good fortune to be exposed to math teaching at school by an extraordinary teacher when I was sixteen. This man was not a professional teacher but a scientist whose research institute had ceased operating because of the war. He taught as a sort of hobby or pastime until he could resume his career, and fortunately for me, he created his own curriculum. I thus escaped the boredom of run-of-the-mill arithmetics teaching.

I really shouldn't have used the word "qualifications" as though one had to be certified by the mathematical establishment before writing a popular mathematics book. Mathematics doesn't "belong" to anyone, let alone to the mathematicians.

Enzensberger addresses the paradox that vast populations are afflicted with mathophobia despite the fact that civilization has been increasingly mathematized. Their ignorance and indifference are profound. Where to place the blame? What can be done to alleviate the affliction? The obvious targets for blame are the mathematicians themselves, the teachers of mathematics and the curricula they follow, particularly in the elementary grades, and the expositors of mathematics, including science journalists and the public relations releases of mathematical societies. There is, of course, overlap in these groups. The blame is less often placed on the fact that the material itself, of its own nature, may be difficult.

Enzensberger finds the mathematicians guilty of elitism; the teachers and their curricula guilty of deadly dullness; the expositors, if professional mathematicians, guilty of stiff-neckedness and an inability to lessen precision of statement in order to achieve increased comprehensibility. Professional journalists, of course, do the best they can with the official news releases and try to talk the public into excitement over whatever it is that the mathematics establishment deems is their latest blockbuster.

What is Enzensberger's solution? Convey the poetic, imaginative stuff of mathematics; leave the rote behind whether in arithmetic or in more advanced topics. Tedious rote does not reveal the real substance of mathematics even though the public may think it does. And leave technical reasoning behind.

While admitting the wonder or the miracle that "something akin to *l'art pour l'art* (art for its own sake) should be so capable of explaining and manipulating the real world around us," in the tension between the esthetic and the utilitarian aspects of mathematics, Enzensberger definitely leans towards the former. His recommendation, hardly unique, is to start children in by arousing their wonder, their sense of mystery; interest in the bread-and-butter stuff of mathematics will follow. Interest in the details of our mathematized civilization will follow.

Perhaps.

As regards the communications gap, there is no substantial way of bridging it; and I will demonstrate this by a physical analogy. I watch an exhibition of figure skating. What the skaters do is difficult and beautiful to behold. I know that there is no law of physics that would prevent you or me from going out on the ice and doing likewise. But this is impossible; it would take us years of specialized training, not to mention native ability.

Though mathematics is largely mental, I believe the analogy carries over. The best we can aim for is to enlarge that public that appreciates how it is affected by mathematics.

Part VI

At Brown University

My Problem, Dear Ann

My problem, Dear Ann, was that, like many people, I wanted to pursue several incompatible careers simultaneously. I wanted to do research in mathematics and I wanted to write. The National Bureau of Standards was an excellent place for the former. It was not for the latter. No government agency is; if you want to write, that's your privilege, but not there. You'd best get out.

While in the early Fifties it was fairly difficult for a new Ph.D. to get a "dream job," the early Sixties were possibly a "golden age" of science employment in America: jobs galore. University research funding by the government and private foundations was at a high level. This was due, in part, to the launching of Sputnik I by the Soviets in 1957. The technological and defense communities in the USA were in momentary shock. Although they knew in principle what was going on in space development, they had not anticipated that an actual space shot would take place so soon.

The response was fast and splendid. Money, effort, and talent poured into space projects and into the whole of scientific and technological research and development. As befitting its shift in emphasis from aeronautics to space, my old employer, the NACA, changed its name to NASA. Since its inception in the early Twenties, the agency had always had a strong lobby in Washington and was popular with Congress; its pockets were very deep indeed, and it became a funding agency in its own right for projects in universities.

Once I had made up my mind to leave Washington, and made this known to professional colleagues, I received numerous offers. They came to me, not I to them. There was no need then as there is now, to prepare a dossier in 163 pages and send them out in hard copy, e-mail, or both, to 163 department

chairs. No necessity to prepare a home web page whose overblown self-adver-
tisements might serve as a model for aspiring young politicians.

I responded to a bid from Brown University, top-notch in applied mathematics, and located in my native New England. In my letter outlining my
plans, I wrote that I wanted both to do research and to write. Initially the
department believed the former, but not the latter; and despite the fact that
writing takes up as much time and thought as research (though of a totally
different sort) and writing comes with no huge departmental overhead allowances, I was kept on for three decades and, I like to think, to the satisfaction of
all parties concerned.

In my years at Brown, there have been great changes: in the Division of
Applied Mathematics, in the student body, and in the social and technological world in which we are all embedded. Studies of fluid and solid mechanics,
which dominated the division when I came aboard, have given way somewhat and there has been great emphasis placed on dynamical systems.
Probabalistic or stochastic approaches have increased by being applied to
such areas as pattern recognition. Numerical analysis (a term that didn't exist
until George Forsythe coined it in the early Fifties) has become of prime
importance throughout science, a good deal of it "chipified" and hence hidden from open view. Computer science, which in its earlier days was cradled
in my division, declared its independence, and responding to the tremendous
advances in hardware and software, has grown internationally into a technology, a methodology, a metaphysics and a business through which almost every
aspect of life must now be threaded.

In the philosophical aspects of science, it was a period when the idea of a
"universal law" or a "theory" became old fashioned and was slowly down-
graded to that of a "model," a term that stresses the transient, provisional, or
relativistic aspects of mathematical applications.

During my years at Brown the country at large was dominated by the
politics and economics of the Cold War. These years also saw ethnic riots, the
amelioration, actual or attempted, of prejudicial imbalances, the disastrous
Vietnam War and the subsequent student protests, and the growth of women's
lib. In universities and schools, it was a period when teachers were simultaneously regarded as television gurus and as ineffective nerds who didn't have
the guts to go out into the so-called real world and make millions. Throughout
the country there was a universal and almost obscene desire to make money
and to make it fast. A tiny minority of well-educated people (usually young)
retreated in protest to the woods of Maine, Colorado or Montana and became

neo-Thoreauvians. Green movements intensified. The increased materialism of life and the resistence to it fed back into the schools and universities in several ways. In the Humanities, "feeling" and "community values" increasingly won out over "knowledge," and courses were given at which the older generation of educators raised their eyebrows and wagged their heads in unbelief. In science, especially at the graduate level, there was a gradual withdrawal of native-born students from such studies to enter the world of business or of litigation.

Throughout this period, I have taught undergraduate and graduate courses in applied mathematics, modes-of-thought courses that combined the history and philosophy of science, computer courses, computer-art courses, and seminars. I have met with some successes and some failures. I have for the most part received collegial support. I have met some collegial opposition derived from what I considered stubborn professional conservatism. I have met with student enthusiasm, student indifference, and student boredom.

Over these years, I have also pursued my research interests, have written technical articles and books and written popularizations of several sorts, even fiction, and have been able to go public with personal views of mathematical philosophy and education that have run counter to established values. If there is a fundamental incompatibility in both time and spirit between research, teaching and writing, my experience at Brown has been a lucky one, supported generously by Brown, and has been as good a compromise as I could have expected in a world of imperfect tradeoffs.

Barney Keeney and
the Lonesome Pine Foundation

The man who gave the final approval to my position at Brown was its president, Barnaby C. Keeney. At my very first interview he asked me how many children I had. I answered, "Four." He asked me their ages. I answered, "The oldest, thirteen; the youngest, three." Barney groaned. In those days, Brown had a liberal policy of paying college tuition costs for its tenured faculty. Nonetheless, he approved of my being hired.

Barney, an Oregonian by birth, a student of medieval English history, a Purple Heart recipient in World War II, had a sharp, acerbic tongue. He was a bit of a cynic. He enjoyed scatological jokes (which I do not). He used to say that he would be the last president of Brown to know each senior member of the faculty personally.

I had hardly unpacked my books and settled into my new office when I perceived that the Division of Applied Mathematics was in the middle of a crisis that was about to become a blowup. Two factions had arisen within the division, call them the P's and the R's, which were at daggers drawn. As a result, the division had become a hornets' nest of rumor, intrigue, personal recrimination, and threats to leave.

Within a few days of my arrival, I was solicited by representatives of both the P's and the R's, and asked to join their cause. Saying that I knew nothing of the matter (and even today I know very little about it), I refused to associate myself with either party in their cause. Socially, I was at ease with both.

The P's and the R's brought their complaints to President Keeney. Keeney responded by saying, in effect, "a plague on both your houses, I will not respond to threats. Settle the matter among yourselves." And settle it they did: the zealots of both parties ultimately left and took jobs elsewhere.

It took several years after I arrived before this scenario played out fully. I wondered at the time what might be the deeper roots of the quarrel apart from the complaints that were aired in public. Slights? Personal animosities? Jealousies? Cravings for academic power? The enervation of the scientific fields of the P's and the R's ?

Whatever it was, it was clear that what fanned the blaze was the availability of academic jobs. A confluence of events, the baby boom, the political and social ideal of near universal college education, the Cold War and Sputnik, the discovery of new technologies, including the digital computer, had created university positions galore.

The star system—always with us as a psychological necessity but more dormant in some periods than in others—burnt fiercely. It raised salaries and elevated jealousies and egos. New colleges were inaugurated, old teachers' training schools suddenly were raised in status to universities and Ph.D. programs put in. Everywhere department chairmen tried to populate their departments with research people of the first magnitude. Why should they not? Wasn't this Jefferson's aspiration for the University of Viriginia?

Professorial chairs named after wealthy donors sprung up like mushroons after a warm, wet, and foggy night. They were used as carrots, and a Professor of Inner and Outer Tangentialism hardly rated unless he (later: or she) was also the Bumble Bee Honey Professor of the same subject, generously endowed by the will of B. Longworthy Bumble.

If a person was unhappy at University A, all he had to do was to leave (or threaten to leave) because he had a firm offer from University B at 1.5 to 2.0 times the salary he was receiving. Departmental blow-ups occurred with a higher frequency than Hollywood divorces.

For some years I profited from the system of contracts and grants (I had grants from the Office of Naval Research and also a Guggenheim Fellowship). Though I realized what a great stimulus these were to science and technology, I did not like the system and I got out of it as soon as I thought I had mustered sufficient academic independence. This defiance very likely made my department chairmen (later: chairpersons) unhappy because it deprived the department of the substantial overhead that could be charged to each grant.

After a few months at Brown, thinking about the entire system of grants and contracts, private, governmental or industrial, I vented my displeasure with the system (particularly with private foundational grants) by writing a short satire called "The Lonesome Pine Foundation." It was printed in the August 1964 issue of *Harper's* and represented my first appearance in the so-called slick magazines. I found my words cheek to jowl with those of well known writers as Ralph Ellison, John Kenneth Galbraith, Joseph Kraft, and Luigi Barzini.

My opinion of "grantsmen" and the benefactions they created was lower than the dust. I saw millions, no, billions of dollars being thrown down the toilet. In "The Lonesome Pine Foundation," I dreamed up an imaginary foundation, located in Bangor, Maine, endowed by a wealthy eccentric, run by myself, and totally removed from pressures that have come to be called political or technological correctness. Zachariah Smith, who had made a fortune by selling pine-needle sachet pillows, had given me $1,250,000 a year to distribute to worthy applicants. So you see that as foundations go, it was a small one. I described briefly a number of grants made by the Pine Tree Foundation. A couple of them have proved prophetic.

My term as chief distribution officer of the Pine Tree Foundation came to a dramatic end. Mr. Shingle, a lawyer for one of the large Maine lumber companies and a trustee of the foundation learned, somehow, that I had managed to siphon off a substantial sum to an applicant called the Fig Leaf Fund of which, it turned out, I was the trustee and executive director.

I was shown the door, but the grant to Fig Leaf remained intact and I moved off to the hollows of Vermont, reduced in cash but enjoying the freedoms of the Shingle-free environment.

President Barney Keeney happened to read my satire and liked it. He buttonholed me after a faculty meeting, told me so, and then related a story (in unrepeatable language) of some stupid foundation with which he had recently locked horns. Not long after, he left Brown. He said that a ten years' (1955–1966) hitch at the job was enough for a person. On the side, he had been instrumental in creating the National Endowment for the Humanities, and accepted a job as its first head.

I saw Barney one more time in his office on G Steet, in Washington, but failed to ask him whether my Lonesome Pine article was affecting his decisions. The "Endowment," as it is called familiarly, and its sister, the National Endowment for the Arts, have in recent years been in serious trouble with Congress for sponsoring material that would much more appropriately have been covered by the Fig Leaf Foundation.

Technology: A Light That's Failing?

I have been reading that in contrast to the post-World War I years, native American students have been opting out of mathematics and science. I know from personal experience that the ranks of Brown University graduate students in applied mathematics now consist in a much larger measure than ever of non-American, non-Western European students. And I have learned that something similar is true in England and in Germany.

Why? The end of the cold war? The fact that science is both hard work and is poorly paid compared to other professions? The feeling that the world is beyond redemption so let's just make money however we can and get our kicks while we can? The prediction that a meteor is scheduled to plunge into us in 2028, and if this date is miscalculated then there will surely be one at a later date?

In the fall of 1963 I was hardly settled at Brown when I was asked by one of the undergraduate societies to give an evening talk on a subject of my choice. I chose as my title: "Technology: A Light That's Failing?" I think I was reacting to the Cuban Missile Crisis of October, 1962.

Ever since Prometheus stole fire from the gods and was punished severely, people have been talking about the downside of technology. In the post-Promethean era, the most prominent critic of the modern era is Ned Ludd, who in the late eighteenth century destroyed mechanized stocking frames. The "dark satanic mills" of William Blake became a synonym for technology.

I'm not sure what I wanted to get across in my talk; it was probably an admonition that the audience rid themselves of the notion that mathematics is a morally and ethically neutral subject. Mathematics, I told the young students, doesn't exist apart from the people who create it, teach it, and use it and it makes for a certain mind set that humanistically speaking has plusses and minuses.

Today, scarcely a week goes by without someone publishing a book on the downside of technology (but hardly yet of mathematics). If I wrote such a book, it would probably bear the title *I'm Not a Luddite, but...* I have a good friend whose title would reverse the emphasis: *I am a Luddite but How I Love the Internet*.

Recent awareness of the negative aspects of technology abound. They run from the picketing of atomic power installations and the concern whether proximity to high voltage lines induces cancer to whether or not we should be canning and eating peaches out of season. The invasion of computer systems by hackers, now a criminal offense, and letter-bombing terrorism can be derived from Luddite impulses.

The public today is of two minds. It is suspicious of technology even as it revs up its consumption of technological products. By way of compensation or amelioration, if one can call it that, a new book is published every week on finding God through science or through Gödel's theorem. Perhaps one of the authors was in my audience in October, 1963.

Left Lib, Right Lib

In the early Sixties, special interest groups, large and small, were pressing their particular claims in court. The notion of "rights" seemed to overtake the notions of "obligations" and "personal responsibility." The class action suit emerged as a universal procedural mode wherein rights could be enforced and injuries compensated by the award of damages to a vast number of people.

Some of these suits reported in the papers struck me as petty; my reaction to them was to write a spoof. I fantasized that a certain Bernstein, who was naturally left-handed, felt aggrieved because automobile traffic in the United States proceeded on the right side of the road, and that his "adherent tendencies" were disadvantaged by the present road system. He sued in the Kings District Court, State of New York, claiming that the equal protection clause of the Constitution had been violated. Various groups immediately came forward, pro- and anti-Bernstein, filing briefs, *amici curiae*, and the case worked its way upwards to the United State Supreme Court.

Having written the article and titling it "Left Lib, Right Lib," I wondered what, if anything I should do with it. I had had one of my articles published in Harper's Magazine a year or so previously, and I thought it would be nice if I could place "Left Lib, Right Lib" similarly. On rereading what I had written, I detected a slight politically conservative tinge to it, and so I decided to send it to the *National Review*, run by William F. Buckley, Jr., which at that time was probably the most prestigious conservative journal in the country. Within a few weeks I received a short note from Mr. Buckley himself accepting my article, and in due course it appeared in print.

There is a bit of a mathematical background to my having selected right-hand traffic as the subject of a lawsuit. Right-handed or left-handed systems in mathematics are known as "orientations." Though the selection of an orientation is perfectly arbitrary, most geometrical systems carry right-handed, counter-clockwise orientations. That the established orientation of traffic is arbitrary is borne out by the fact that in the U. K., in some of its former colonies, in Japan, and some years ago, in Sweden, one drove on the left side of the road.

The question of orientation also appears in theoretical physics. One of the basic assumptions in many developments, is that the universe is *isotropic*, that is, there is no preferred direction as regards what is happening. In the early Sixties, this hypothesis was shown to be incorrect in certain quantum-theoretical situations. This work by Yang and Lee won a Nobel Prize in 1957, and I suppose it was on my mind at the time I wrote *"Left Lib, Right Lib."*

I collected my $25 or whatever from Mr. Buckley, and proceeded to forget about the whole matter. About a year or two after the publication of the article, I got a long-distance call from a woman who said she was a researcher for Columbia Broadcasting Television. She told me that she was working up material for a series dealing with the treatment by the public of physical disabilities and could she ask me a few in-depth questions about the Bernstein Case described in "Left Lib, Right Lib."

Where had she learned of my article, I asked her. Well, it was listed in some index. Had she read my article? Her answer, as I recall, was ambiguous. If she had read the article, I thought, she would have found the satire laid on so thick, the basic situation so preposterous, the names of the judges so checkably false, that she would have recognized it immediately as a spoof. I concluded that she had not read the article.

At this point I was faced with a dilemma that required resolving within seconds. Should I or should I not go forward with the spoof? Yes? No?

I decided against going forward. I admitted to my caller that I had written a spoof, and that I would have supposed that it would have been referenced in the annals of humor rather than those of law. I could hear in her reaction that she was hurt and disappointed.

I learned a great truth from this incident: people are easily taken in. Civilization proceeds more on the basis of faith than we are apt to admit. Indexes (now called databases) split and give birth to other indexes. That the world of communication abounds in misinformation whose quality is never judged by the laity or is essentially impossible for them to judge. Who knows the extent to which Bernstein vs. the State of New York may now have diffused into hundreds of legal databases.

And it is not the laity alone who cannot judge the material. Let me move forward a third of a century to the Sokal Affair. In an 1996 article in *Lingua*

Franca, an American academic journal, Alan Sokal, a professor of physics at New York University, wrote

> For some years I've been troubled by an apparent decline in the standards of intellectual rigor in certain precincts of the American academic humanities. But I'm a mere physicist; If I find myself unable to make head or tail of *jouissance* and *différance*, perhaps that reflects my own inadequacy.
>
> So, to test the prevailing intellectual standards, I decided to try...an experiment: would the leading North American journal of cultural studies publish an article consisting of utter nonsense if (a) it sounded good and (b) it flattered the editors' ideological preconceptions?
>
> The answer, unfortunately, is yes. Interested readers can find my article "Towards a Transformative Hermeneutics of Quantum Gravity" (!), in the spring 1996 issue of *Social Text*....
>
> What's going on here? Could the editors really not have realized that my article was a parody?

Sokal's spoof has by now been mulled over from many points of view. The question of what is or isn't nonsense is not so easily decided. For this I have the word of a professional philosopher, as the reader will see in a later section that brings in Freddie Ayer.

Opinions of the Famous:
Otto E. Neugebauer

I met him first in the fall of 1963 when I joined the faculty of Brown University. For many years, prior to his return to the Institute for Advanced Study at Princeton in 1984, I ate lunch with him fairly often in the Brown cafeteria known as the Ivy Room. We were not infrequently joined by other members of the history of mathematics department or their visitors. The conversation was relaxed but lively, scholarly but usually very general, and was terminated promptly when the last person had finished lunch. Neugebauer was not one to twiddle his spoon leisurely in a second cup of coffee.

Otto Neugebauer (1899–1990), preeminent scholar in the history of ancient science and mathematics, had that soft appearance and low-decibel manner that I associate with Austrians. Inwardly, he held firm opinions and prejudices, and would occasionally burst out in anger and irritation using English swear words that one felt were an uncomfortable translation from German originals. Not unlike Mark Twain, he was a misanthrope; he perceived the human world as consisting largely of fools, knaves, and dupes; and when he was overwhelmed by this perception he took refuge in his love of animals, which was tender and deep. I recall his telling us (when Hadassah and I visited him at his daughter's summer cabin on Deer Isle, Maine), how he had made a pet of a fox who came visiting for food. On Deer Isle, walking around dressed comfortably in a pair of overalls he had picked up at Epstein's, a general store in the village, he did not play the great professor.

He had his roster of The Greats in his profession, and although his ratings were not as finely tuned as those of G. H. Hardy, anyone who ate lunch with him would find out within a week which of the great names were really great and which were intellectual asses.

Otto E. Neugebauer.

As regards the past, he thought that Copernicus was overrated—he called him Koppernickel. Kepler was much better, and he loved Arthur Koestler's popularization of Kepler in *The Sleepwalkers*. Claudius Ptolemy was a great hero. As regards contemporaries, he expressed his views candidly. I was occasionally shocked and have no desire to go public with them.

Neugebauer could not abide philosophy; he thought it a great waste of time and rarely discussed it. I would guess, though, that his largely unspoken philosophy of science was that of the logical positivism promulgated by the Vienna Circle in the 1920s. He parodied Hegel. Born into a Protestant family in Catholic Austria, he could not abide religious ritual or theological dogmas, and though he was a considerable student of these matters, he parodied them all mercilessly. As a relaxation, he would read *The Lives of the Ethiopic Saints*, and made a dossier of the names of the individual devils that those saints had to contend with.

He was given to playful irony and loved Anatole France's ironic fantasies. I recall his being greatly amused by a passage he had read in an ancient Near Eastern medical text that said crocodile droppings were prescribed for such and such a medical condition. A patient applied them without success and went back to the practitioner with a complaint. The practitioner asked, "What was the sex of the crocodile?"

"A male, I think."

"Then try a female."

On this story Neugebauer commented with a twinkle in his eyes, "You see, the ancients knew all about hormones!"

The work that made Neugebauer world famous was the reconstruction of, the understanding of, and the interpretation of ancient scientific texts, Egyptian, Babylonian, Ethiopian. He led us to understand the rich body of mathematics created by the pre-Greek Babylonians, whose general civilization he found very uncomfortable. The text was the thing. Correspondingly, I would guess that his philosophy of history, if he ever expressed it openly, would have been that of Ranke: "*die Vergangenheit wie sie eigentlich war.*" (The reconstruction of the past as it really was.)

I once asked him to comment on the assertion that one often sees in print that the stones in Stonehenge, England, were placed so as to line up in this, that, and the other manner with the various celestial bodies, and that the field of "paleoastronomy" as it is called was burgeoning. His reaction was short and pointed: "*Quatsch.* Fiction."

I suspect that his philosophy of mathematics (rarely made explicit) was that mathematics exists and has a reality independently of the people who created it. After all, for a number of years, he was associated with Richard Courant (a great friend) and the Goettingen group. I could see at the very beginning of our friendship that his platonic idealism was far removed from my philosophy of mathematics, and in the interests of that friendship, I stayed away from the topic at lunch time.

In point of fact, he perceived the tension between the underlying individual culture and the universal aspect of mathematics; he admired the writings of the famous Egyptologist Kurt Sethe on these matters, but when the chips were down, according to a splendid biographical memoir written by Noel Swerdlow, "even through years of allowing that mathematics was grounded in culture, he really believed that in a more profound sense it was not".

Deep down, then, he was a Platonist; accordingly he was suspicious of attempts to link the history of mathematics to general history. He relegated such attempts to the world of the unprovable, at best, or to the world of fiction, at worst. The text, the symbol, and the recoverable platonic meaning thereof: these were his touchstones and his standards.

Neugebauer's attitude (I should not call it philosophy) towards education comforted me. I always considered myself a bad teacher, particularly with thesis students, thinking that students should be given a small push at the beginning and then encouraged (or even required) to go their own way. This has not been the practice of some of my colleagues. If Neugebauer had a philosophy of education, then according to Swerdlow, this was it, both for himself and for his students, and he justified it satirically by asserting that "no one had yet invented a system of education that was capable of ruining everyone."

At the time I met him, he had for some years been the center of a large international "court" of scholars, students, associates, visitors. He was the Sun and they were the planets, and I sometimes felt that the gravitational attraction experienced by the planets was too great. But zoology and not astronomy was the metaphor used by the court to refer to themselves. He was the Elephant, of course, and there was a Rabbit, an Owl, a Squirrel, etc. I was never a member of this court, only a hanger-on, and if he ever referred to me zoologically, I was not aware of it.

One spring after Neugebauer's death, I was in Copenhagen and looked up Professor Olaf Schmidt, his former Ph.D. student and great admirer. Schmidt picked me up at my apartment in his car and then proceeded to give me a Neugebauer Tour (I was going to write "A Tour of the Stations of the Cross," but realized the inappropriateness of the metaphor) of the sites in Copenhagen that were connected in some way with Neugebauer's exile years in Copenhagen (1934–39). As a Professor in Goettingen Neugebauer had not been threatened personally by the racial laws promulgated by Nazi Germany, but because of his opposition to them, he was regarded as *politisch unzuverlässig* (politically unreliable) and he had the sense to get out of the country. He knew very well that history was more than text, that it could not be reduced to a few symbols. He once remarked to us: "If you never heard the sound of Nazi boots below you in the street, you cannot understand the history of the period."

In one of his prefaces, Neugebauer thanked the ghosts of ancient Babylonian scribes who "by their untiring efforts they built the foundations for the understanding of the laws of nature which our generation is applying so successfully to the destruction of civilization".

At least, he added, "they also provided hours of peace for those who attempted to decode their lines of thought two thousand years later".

I'm sure he would have agreed with Shakespeare (in *Macbeth*), that human history "is a tale told by an idiot, full of sound and fury signifying nothing." The very separation he was able to make between the human world and the world of platonic ideas was a source of strength, a basis for his faith and for his ability to pursue his work with gusto to an advanced age.

Warm Bodies and Cold Calipers

Sometime in the mid-Sixties, I was encouraged to take up oil painting as a hobby. I took to it, taught myself a few things, painted on canvas, canvas board, wooden shingles, or whatever plane surface happened to be around. I had fun and relaxation. I threw out most of what I produced and retained only a few of the better ones. Oils, I found, are much easier than water colors. If they weren't so expensive, they should be given to children. Acrylics (which were hardly around when I started to paint) are more difficult than oils. The main trouble with oils is that they dry so slowly, but this is also a virtue for it means that areas can be rubbed out and repainted without too much difficulty. I read somewhere that the great Turner used to rub out and repaint even after his work was hanging in the galleries.

Feeling strengthened by my success in producing some decent things, I thought I would seek professional instruction and I enrolled in a night class in oil painting. I thought that the instructor would tell me what sort of brushes to buy, how to hold them, how to apply the paint, in short, all the mechanical aspects of the craft. I thought the instructor would set up his own easel and work away in demonstration while the students watched. Nothing doing. That kind of instruction was for the TV art classes that peddled kitsch.

I came to class with my equipment as required. Our instructor was a teacher at the Rhode Island School of Design (RISD), whose name was Italo Scanga. (He is now in La Jolla, University of California, and in the intervening years has had many successes of various kinds). He set up a still life consisting of an arrangement of commonplace objects. He told us simply to get to work. So we got to work, about fifteen of us. No starting off with "spots and dots" exercises.

Scanga walked around the studio from student to student. He liked this, he liked that. He didn't like this over there. Put a little more joy into your work, he would say, or, that's nice. I recall him as a generous, encouraging critic.

In this way, the class proceeded for about a month doing the still lifes Scanga set up. As I worked, I also walked around from easel to easel, (as did all the students), looking and exchanging information and opinions. It amazed me to see the variety that was inspired by one and the same arrangement of objects. As regards my own talent, I judged it to be right down the middle when I compared it with that of my fellow students.

Then came the day when Scanga told us that next week we were to bring in materials for our own still life. And make them unusual, he added. I brought in a bunch of broccoli and a canvas board about five feet long and two feet wide. I intended to fill the board with a heroic broccoli, a proud and triumphant broccoli, a broccoli rampant, a broccoli inspired more by the equestrian statue of Marcus Aurelius than by van Gogh's pair of shoes. And I went to work.

I painted a yellow background for my broccoli, and then sketched an elongated outline of my vegetable in such a way that it filled eighty percent of the canvas board. Then I proceeded slowly with solid modeling (three-dimensional) using many shades and intensities of green that blended into one another. I stressed the articulation of the stems and shoots and leaves and flowerets. It took me two sessions and I wound up with a grandiose and haughty broccoli, a broccoli with considerable backbone.

Italo Scanga came around and looked at my oeuvre. "That's nice," he said. And I felt a million feet tall.

The term was getting on; the snow was beginning to fall. The studio, an old and essentially abandoned laboratory from the days when old fashioned mechanical engineering was taught to undergraduates, was cold and filled with discarded and rusting gears, generators, and power transmission belts. Our instructor now decided that it was time that the class graduated from still life to human life. Next week, we'll have a model, if she's over her cold. We'll have her for three times. Everybody bring in $5 to pay for the model. This was a first for me. I didn't know what to expect.

I brought in a 2′ by 3′ canvas board and my $5. As promised, when I arrived, the model was already in the studio sitting on a slight platform in a robe. She was no Botticelli Venus rising from the surf; she was a very plain lady who had made her living by posing for RISD students over several decades.

"This is what the model will do tonight," Italo Scanga told the class. "She'll stand up and hold a pose for five minutes. Then she'll rest for five or ten. Then she'll resume the pose or a slightly different one, alternating in this way for the whole period. So get going and arrange your work accordingly. And work fast."

Get going, indeed. Not a word did Scanga say to us about the strategies of painting a nude. The world of art is full of nudes and we'd all seen them: from the Venus de Milo to Picasso's nudes. Paintings of nudes come in hundreds of sorts, sizes, shapes, positions, arrangements, groupings, classical, mannerist, allegorical, abstract, you name it; the whole history of Western art can be shaped around the fascination of artists with the human, particularly the female, body. The famous art critic Kenneth Clark conjectured in one of his chapters in *The Nude* that the impulse for mathematics with its line and curves came from men's fascination with the female body. Historians of mathematics have not elaborated or even mentioned Clark's conjecture, probably for fear of being labelled "sexist."

Scanga's injunction to get going in the presence of a real live nude was like tossing an infant into the raging sea and saying to it: "Swim." Was Scanga himself capable of painting a nude? He was then in what I called his "cool tube" period: cylindrical sculptures whose surfaces were painted or stained with a slight blend of cool greens, blues, and pinks.

And I? I was still in my broccoli period, inspirationally speaking. The model stood up and faced the class. She doffed her robe, kept it flung over her left arm, her right hand on her right hip. Full frontage as they say now in the movies. My eyes blurred and my head whirred.

Where the model had a torso, I saw a green stem; where she had arms, I saw green shoots; where she had breasts, I saw green flowerets. In my mind, I converted the model into a broccoli and I would call my work "*La Récolte*" or some such. In fantasy, I may have made this suppressional transformation, but what emerged on the canvas board after several rest periods was a woman, and recognizably so.

Italo Scanga walked around from student to student commenting. Ultimately he came over to me. He looked, he saw, he pursed his lips. "The arm is out of proportion to the leg. The leg is out of proportion to the head."

I heard his judgment and was confused by it. Wasn't this the age when we accepted that Modigliani painted women with tremendously elongated necks? That Picasso painted women with several noses? Wasn't Chagall still painting women floating in the air with bodies twisted around like pretzels? How does this guy get off telling me that what I had laid down in paint was out of proportion? Perhaps the disproportions were my unconscious artistic intent driving me forward.

I looked at Scanga with blank non-understanding. He repeated what he had said. I continued to stare at him. He went over to his desk and got a huge pair of cold steel calipers, and gave them to me.

"Go over to the model," he said, "and measure her. You'll see."

I hestitated.

"Go. She's used to being measured by students."

I went. I applied cold steel to warm but shivering flesh. I measured. The model stood absolutely unperturbed. Did I know really what I was doing? I doubt it. I went back to my easel and made some computations. I took my rag and wiped out a portion of the figure and redid it. Here, in the flesh, was mathematics applied to art. What is gained, what is lost when this occurs?

My nude remained in my closet at home for a number of years. I was not proud if it. Ultimately, I threw it out.

It was a different story with the broccoli. I was proud of that. A day arrived when my wife and I found it necessary to give a present to our friends the Petersens. At that time they were living a quasi-rural life with vegetable gardens and such and I figured that it would be quite appropriate to give them my broccoli. I told my wife that any original oil is far brighter, far more cheerful, than any reproduction of Monet or of whomever. She agreed. I signed my work (quite modestly) in the lower right corner, wrapped it up, and, at an appropriate moment, gave it to them.

"This is an original oil," I explained. I looked around to see whether I could spot any original oils on the walls. I could not.

"Who is it by?" Arnold Petersen asked me.

"Elève de Scanga," I responded with considerable pride.

"Who is Scanga?"

I told the Petersens about all the glorious successes that Italo Scanga had had. (He was then in Philadelphia.)

"And who is the élève?"

I drew myself up a bit.

"The élève is none other than yours truly. Who else?"

"Well, you don't say so! Who would have thought that?"

Despite these expressions of pleasure and surprise, I think that my broccoli is not now to be reckoned among the extant works of western art. The Arnold Petersens moved back to civilization and when visiting them in town after they moved, I neither saw it on their wall nor heard about it further. I had never taken a photo of it. But my broccoli—my particular broccoli—lives on, I have been assured by some philosophers, in the eternal world of platonic ideas.

The Very Early Days of Computer Art

One of my specialties within mathematics is called "approximation theory." This is where one learns how to represent curves and surfaces by mathematical formulas. When the formulas are computed up and displayed, the representation is often found to be a mathematical idealization or approximation to some physical situation. Hence the designation. Once a mathematical statement of a curve or shape is determined, the shape can be altered or transformed very easily by purely mathematical means. Curves are related to two-dimensional art and surfaces to three-dimensional art (or sculpture), so considering the hundreds of mathematical specialties that exist, mine was one that was reasonably close to art and to commercial design.

Within approximation theory, the characteristics of the curves and surfaces produced are studied in great detail. Since the number of formulaic options is essentially unlimited, the subject of "best approximation" is born, whose duty it is to pick out of the myriad of possibilities, the one that is best according to some criterion given in advance.

In 1913, Serge Bernstein, a young mathematician in Russia proposed an approximation scheme, now known as the Bernstein polynomials. These polynomials have the property that while they mimic given shapes quite well in a gross sense, they do not get close to the given shape in an absolute sense. I knew about Bernstein's work, liked it, and in a book I wrote on approximation expressed my opinion that Bernstein's polynomials might find great use in design work.

This throwaway remark proved prophetic; it was picked up by researchers, some of whom had been my students, Bill Gordon and Bob Barnhill, for example, with the effect that a generation later, Bernstein polynomials are a fundamental and standard concept in a new field called computer-aided geometric design. Spline curves, whose theory was deepened during World War II by I. J. Schoenberg, have also played a prominent role in design work.

These mathematical objects have been generalized, abstracted, reworked, rediscovered, improved, and applied. Newly designed auto or airplane surfaces might now have a Bernstein basis or something akin underlying them, as might the solid objects depicted in virtual reality landscapes.

After obtaining his doctorate in applied mathematics, Bill Gordon left Brown (in the late Sixties) and shortly thereafter found himself employed in the research laboratories of the General Motors Corporation in Michigan. Some months later, he was able to invite me to spend a week as a mathematical consultant. This was in the glory days of General Motors. The research labs were located in a beautiful complex designed by Eero Saarinen. The company was driven by its sales and body design departments. Mathematicians stood pretty low on the totem pole.

I was shown around by a minor executive driving a company car which, essentially, he treated as his own. I asked him whether they changed his car when the new models came out. He answered that they changed his car whenever the ashtray got full.

Although design, research, and development was company confidential, I managed to see some of the standard graphical or drafting techniques employed by the automotive industry: huge plotters, driven by early model computers, capable of producing full scale drawings.

When I returned to home base—I was then teaching the subject of interpolation and approximation in a very theoretical manner—I set as a problem to my students the design of a font of English letters by means of periodic, interpolatory splines. The students went after this assignment with gusto and every alphabet they created had an individual character. By putting the individual letters together into words, I began to appreciate the problems of style and legibility that the great typographic designers faced and solved. If I had pushed this research into its commercial possibilities, as did other people, I might now be very wealthy. But I had other fish to fry.

As a result of these various experiences, I determined (Fall, 1972) to give a course on computer art. This was agreeable to my department. One of my younger colleagues, Charles Strauss, a mathematician and a crackerjack programmer was enthusiastic about the possibility, and we did it jointly. It was one of the first such courses offered in the country. I like to think of computer art as a marriage of art and mathematics.

This course was offered in the days when computers still used punched cards and long before the days of the personal computer, nice, sophisticated paint programs or versatile programs that do computer-aided design, hi-fi digitizers, and high-resolution video displays and printing devices.

Ten students or so registered for Special Themes and Topics 29, about half from art and half from computer science. Students in mathematics and other sciences were just not interested; it was too far out a subject. One of the first things that Charles and I learned was that though both kinds of students spoke the English language, when it came to the goals and methods of the course, there was hardly any communication between them. In time, we were able to overcome this, but not entirely.

We brought in guest faculty from the art departments of Brown and the Rhode Island School of Design. Same thing: very little mutual understanding or communication. One member of the RISD art faculty went over to a large piece of computer equipment and painted a moustache on its metal housing. "You wanted computer art," he said, "well, here you have it." A joke? A protest? A serious statement? We never found out.

Charles taught the students how to program straight lines, rectangles, circles, other mathematical curves and how to combine them, replicate them and transform them. We were working at that level. Eventually, we sensed that some degree of understanding had been established. Things began to move. Mondrian, Miro, Kandinsky were simulated.

We told the students that at the end of the course we would have a gallery show of their best work. The students went at it with gusto and developed new projects for submission. They were proud to have their worked framed. We booked into the Woods-Gerry Gallery which serves as the students' gallery for the Rhode Island School of Design, announced the show to the public and alerted the art critic of the *Providence Journal*.

What was produced? Lots of geometric art, op art. Collages; some were hand painted. The outline of a man's head, totally contoured internally, some proto-animations: still shots of a human figure in motion. Some transformations: man to woman, elephant into donkey, that sort of thing (very easy to program!) One student submitted a supermarket shopping basket full of crumpled computer output paper filled with code.

I didn't offer the course again, feeling: let others take over. And they did. I went back to creating a few more Ph.D's in traditional applied mathematics. But my interest in art must have spilled over into the theoretical work, as several of my Ph.D students took jobs in companies that were developing computer graphics.

Looking over the old slides of the January, 1973 show and comparing what I see of recent work, I am amazed at what computer technology has accomplished in the last quarter century. Take commercial or industrial art, for example. *IRIS* (the company journal of Silicon Graphics) noted:

> No longer do automakers need to carve huge, life-sized clay models of new cars to wheel into wind tunnels. Gone are the days when whole automobiles had to be sacrificed for crash tests. To-day, it can all happen digitally, in real time and in 3-D.

And it is not just the automobile industry that has profited from computer graphics, but the whole of science and technology, from aerodynamics through medicine and all the way to zoology.

Nonetheless, the principal philosophical problem abides and perhaps is intensified: what is art; and what, for the individual artist-programmer, is worth doing? Amateurs may not have to earn their daily bread, but the problem is there for them as well.

A New Esthetic?

Critical discussions of works of graphic art do not, for the most part, treat the details of the physical means by which the art was created. One would have to dig a bit before one found a discussion of how Caravaggio ground his colors or arranged his palette, or how Michaelangelo had his chisels sharpened. The discussions tend to be historic or mythologic or symbolic, allegorical, aesthetic or sociological, biographical, psychological, introspective, religious, ethical, political, deconstructive, in short, anything but technical. And yet, it is clear that the means affects the ends. One has only to look at Asian wall scrolls, produced with brushes and inks that differ from those of the West, or look at etchings of the seventeenth century or look at contemporary commercial art produced with an air brush in order to be convinced of this fact.

The digital computer, programmed in the service of art, represents a revolutionary new artistic means. It is so new (a half century, let us say, which is as nothing in the history of art), that only now are there emerging critical discussions of the medium in its relation to instances of finished products, and of the finished products as art objects to be viewed independently of the means. The craft has developed so rapidly and has delivered such miraculous effects that its practitioners have been held in a state of high elation induced by the available hardware/software, a state of "hoo hah" or "look, Ma, ain't it great what we've been able to do?" Can the elation last? The history of technology shows that the miraculous is often transformed rapidly into the everyday, the ordinary, the expected, and finally into the banal. Da Vinci may have found his final resting place on T-shirts.

One needs materials to convert into art objects: paint and canvas, paper and ink, clay, marble, found objects. But one can create works of art without tools using only the hands (unless one regards, as is not usually done, the fingers, hands, and fist as tools). The artist has always had tools, and among them are some that might be termed mathematical: stretched ropes and rulers, dividers, compasses, grids, splines and templets of various sorts, pantographs, camera obscuras, moulds and shaping instruments. The ancient potter's wheel might even be regarded as an analog device for creating surfaces of revolution. Jacquard looms and automatic lathes programmed mechanically have mathematical aspects that predate digital computers.

The digital computer is an instrument for producing art in which the mathematical idea or theory and the mathematical vocabulary merge much more intimately with the production; in which a simple text, say, "cat," can be a tool, a final visual production, or one ready for further manipulation. So far, the verbal and the mathematical means, the former readily available for inspection and the latter usually submerged and hidden from view, have controlled the ends much more than the visual aspect has. Computer art is a case both of a plethora of tools in search of the artist, and the artist-programmer constantly searching out the possibilities of new tool creation. The computer art tool kit cries out: come, use me, create with me, even as the blank canvas traditionally challenged the artist to fill it with something that had significance.

How many computerized art tools are there? As many as one cares to program; and the tools develop imperceptibly along with what they create, gaining ascendancy as they themselves become pieces of meta-art. The craft still stands in danger of fulfilling the perception of Marshall McLuhan that the medium (or the tool) becomes the message.

One may ask what new elements computer programming brought to add to the capabilities of artists. Collage was around, overlays and the fusing of one image with another can be done quite nicely by photographic means. If the imagination of artists had been so stimulated, it might have produced, with pencil or in oils, fractal images. Look at the inner complexity of paisley fabric designs, for example. This did not happen with fractals; it took the whole thrust of mathematics and of mathematical physics to elicit fractal images and then to propose them as art objects. Furthermore, as the dialecticians assure us, quantity can change into quality, and the rapidity of certain operations, their exactness, their ability to deal with minute detail, result in computer art lending a totally new quality to old operations.

Nihil ex nihilo fit; nothing comes out of nothing. Computer art has at its disposal the whole of the previous experience of art, sculpture, architecture, and photography to build on, and it is not difficult to find specific pre-computer elements that in retrospect seem to anticipate some of the modalities that have surfaced in computer art. The idea of the subdivision of the field into minute areas is present as a technique in the stylized art of ancient Egypt. The pointillism of the 1890s seems to presage the division of the computer screen into, say, 1024 by 1024 pixels, each of which can be addressed, colored, and manipulated geometrically or otherwise on an individual basis.

"Typewriter art," consisting of images created with *x*'s or any of the typewriter keys, date from the early days of the typewriter. It is a form of pointillism, as were the ultimate constituents of the crude photographs in newspapers, which, viewed under a magnifying glass, yielded their atomic dots and spots. This kind of thing extended into the first generation of computers.

Abstract art, for example Brancusi's birds (which were denied by him as abstract), op art, aleatory art (Jackson Pollak dripping paint directly on the canvas), assemblages of found objects (the collages of Arcimboldo, 1527–1593, who assembled vegetables into human or allegorical figures), all predate the computer. The idea of self-regeneration that appears in fractal art is already present mathematically in Bernoulli's spirals and in some of the outlandish decorative excrescences piled upon excrescences of the earlier Mannerist and Baroque periods. Strange geometric transformations of realistic figures were done and the process was called anamorphosis. (Example: the skull in Holbein's "The Ambassadors.") *Trompe l'oeil* is early virtual reality.

What is new in computer art, apart from the manner of production, appears not so much in two- or three-dimensional work but in multi-media, an area that mixes static graphics, animation, video, sound, text and hypertext, and possible interactivity between artist and viewer or manipulator. Even in these striking developments, there are pre-computer precedents. Think of the theater, the circus, the opera, the more recent on-site exhibitions of *son et lumière* for tourists. Passive entertainment has been subject to on-line audience interaction; think of—grisly thought—the thumbs up or thumbs down of a bloodthirsty crowd viewing a gladiatorial show in the Coliseum in ancient Rome.

Art is and always has been related to show, games, religious and secular ritual with their programmatic elements; to education, the necessities of life such as housing and to imagined realities. Computer art picked up on these, moving rapidly into games, entertainment, commercial design, and scientific and technological simulations, experimentation and theorizing. The reaction a few hundred years ago of North European visitors to the lush, sensuous, erotic displays in Italy, anticipates in a primitive way today's reactions to virtual reality.

What is new also about computer art—but not entirely new—is that the art has no locus in a physical object. There is only one Mona Lisa. It is in the

Louvre. It is finished, done with. There may be millions of reproductions, but they are not the Mona Lisa. Nor is anyone invited to add their own few bits to the original Mona Lisa. When a restorer does repair work due to natural decay, some people cry out and lament that the restoration falsifies.

Not so for computer art. Residing somewhere in a computer memory in a dematerialized visual form, it calls out for materialization on the computer screen or on paper, and invites collaborative, interactive accretions.

What have been my personal reactions to computer-generated art? I find that the images of simple things such as bowls are more real than real. The colors more colorful, the images, having crisper definition, are more sharply articulated. There is an exactness of shape that exhibits a superhuman perfection. All is clean, shiny, buffed to a high gloss.

What emotions are stirred up in me? Veneration, awe, fear, love, hate, surprise? Awe sometimes. Occasionally amusement; occasionally disgust when the virtual objects are deliberately vermicular—a genre that seems to appeal to the topologist-computer artist.

On a rather more elevated level, some pieces of computer art have thrown me into the universe of non-accessible horizons depicted in early de Chirico. The gray and difficult world of human troubles is left behind, and I ask myself why should it even continue to exist when a transcendental world seems to be available through technology? A noble thought, but one that probably will crash to the ground as many idealisms have crashed in this century.

While I obtain a feeling of elation in such a world of pure mathematical sensation, there is little in the images that operates for me in a didactic way outside of mathematics itself. This may be contrasted to the religious art of the Renaissance and earlier, when books were far from commonplace, and when art served to tell the Bible stories to draw a moral from them and to provide an inkling of the world to come. At the moment, there is too much imposition of the ugly chaos of contemporary life, of the horror- and fear-inducing images of prehistoric and fictional monstrosities both beloved and dreaded by children. The computer Virgin has not yet appeared to provide peace of mind, comfort and salvation in a horror-ridden world.

Computer art can be inspired directly by mathematics. Mathematical objects often exhibit breathtaking beauty. I am thinking, for example, of spiral shells or of the Gateway Arch in Saint Louis, Missouri (which really should be viewed on the spot and not just on a picture postcard.) There is a kind of computer art that can be described as the making visual of mathematical formulas. In two dimensions this results in an assemblage of points, a curve or a family of curves, or a projection onto a two-dimensional screen of a three or higher dimensional objects. The bare-bones skeletonic object is often filled in or rendered in some manner to simulate solidity.

In actual three dimensions, one has computer sculpture, in which the material is cut or shaped by a computer-driven tool programmed in accordance with a particular mathematical formula. A person who has no mathematical knowledge may view this sort of art and feel the sensuousness of the pure shapes displayed. If the person understands the mathematics that underlies the work, this adds another realm of appreciation.

There was a continued debate throughout the last century about whether, as art changes and provides us with different things, it progresses. The discussions of progress in art are charged with emotion, and can hardly be separated from discussions of the relationship of art and morality.

Technology creates new worlds. As it does so it destroys old worlds. At the technological level, awareness of progress comes fairly easily. With regard to moral progress, one wonders.

What I Learned from Mary Lucy Cartwright

You know, if a person does one great thing in his life, that's enough. If he does two, it's extraordinary. Einstein did two.

At the learning end, if you learn one thing from a person, that's really enough. You don't have to push it. And that's also my philosophy of cook books: one good recipe—a new one—is worth the price of the book.

Apart from lots of in-group gossip, I learned one thing from Mary Cartwright. I first saw her at the International Congress of Mathematicians held at Harvard in 1950. I attended a session of the Congress held in what was then called "The New Lecture Hall," and I asked an older accompanying friend, "Who is that lady sitting next to Norbert Wiener?" His answer was immediate: "Oh, that is Mary Cartwright."

If I now had the guts to turn this true story into a version of the classic joke, I would have written "Who is that man sitting next to Mary Cartwright?"

I was familiar with the name because I had read some of her papers on integral (i.e., entire) functions, the area of my own thesis, and within a few years, I made use of one of her results from 1936. Within a few years also, there would appear two books on entire functions, one by my thesis advisor, Ralph Boas (1954), which contained substantial references to Cartwright's work, and then one by Cartwright herself (1956), both of which I studied carefully.

My intention here is not to provide a description of Cartwright's mathematical work in complex variables and non-linear differential equations; that has been done far better by more qualified people. (See, e.g., the article by Shawnee L. McMurran and James J. Tattersall, on the long mathematical collaboration between Mary Cartwright and J. E. Littlewood, *The American Mathematical Monthly*, Dec. 1996.) My intention is to provide a description of her personality as I experienced it.

Mary Cartwright.

I first met Mary Cartwright socially in the fall of 1968 when she was a visitor at my department at Brown. She was a tall, slender, gray-haired lady and she reminded me (in appearance) of those devoted, spinsterish, mousy, grammar school teachers who had provided me with an elementary education. On further acquaintance mousy? Not on your tintype, as they say.

As the weeks went on, I got to know Mary better, and we dropped the formalities of title. I could see that I was dealing not only with a very sharp mathematical mind (which I already knew, of course), but with a mind and a tongue sharpened over the years by the repartee of Cambridge high table wits. This pleased me greatly, for I can hold my own in the game of academic wisecracks. And so we got on quite splendidly.

In the years that I knew her, I did not engage with her mathematically, though when I came to write the beginning of *The Schwarz Function* (1974), I inserted as an example of the use of conjugate coordinates and in her honor the nine point circle and a story she told me about it.

During her stay in Providence, there was a mathematics meeting in Cincinnati that we both—each unknown to the other—planned to attend. We made separate air and hotel bookings. Surprise: we met on the plane and found out that we were booked into the same hotel. We taxied to the hotel together, I helping her with her bags. Further surprise: the room clerk booked us into adjacent rooms.

Depositing her bags in her room, which was very large and with two queen-sized beds, I decided to tease her.

"Mary, you know if I moved into your room with you, and we split the cost, we would both save a bit of money!"

She was silent for a moment; she considered the possibility, and then she looked at me sternly and said slowly, "I don't think that would do."

At this point, let me interpolate a few facts of the *Who's Who* variety. Mary was born to a Northhamptonshire family where her father was rector in the small village of Aynho. She told us early on that she had lost two brothers in World War I. (I always thought that from time to time a certain sadness would come out from beneath the frequent fun.) She went up to St Hugh's College, Oxford, taking a first in 1923. After a few years of prep school teaching, she was back at Oxford in 1927 to a doctorate under G. H. Hardy in the theory of integral functions.

She was a Fellow of Girton College, Cambridge, from 1930 to 1949 and was its Mistress from 1949 to 1960. She was lecturer in Mathematics in Cambridge University from 1935 to 1959 and reader in the theory of functions from 1959 until she retired in 1968. Among her many honors and awards, she was elected Fellow of the Royal Society in 1947 and was named Dame of the British Empire (D.B.E.) in 1969. All in all, she was a mathematician of world class, and this was a particularly remarkable achievement in a generation when few women entered the field.

Her sense of fun was strong (math was certainly fun) and her dry humor flowed like a stream—to mix a metaphor. The announcement of her being awarded the D.B.E. came while she was at Brown. Eleanor Addison, then administrative assistant to the Division of Applied Mathematics and a woman who takes no guff either from man or beast, told me that when she heard the news, she approached Mary and said to her, "Well, Mary, now that you've been knighted, I suppose I'll have to bow down three times."

And Mary answered, "No. Twice will do."

After Mary left Brown to take up visiting positions at Case Western and Claremont Graduate School, and ultimately to return to Cambridge, there was a gap of a number of years in our friendship. In the mid-Eighties, though, for a week or so in the summer, Hadassah and I would take ourselves to Cambridge, where we had developed a number of friends. Mary was now well into retirement, and in the minds of our Cambridge mathematical friends, she was a local legend and an icon but not an active presence. Our relationship was readily reestablished and we found that in addition to her iconic character (which she enjoyed), she was also quite a social being. And not only a social being, but more than a bit of a gossip.

She had been everywhere worth going to, mathematically speaking. She knew everyone worth knowing in the mathematical world and she ranked them—even as Hardy had done, using the rankings of cricketers. In the larger world also she was a bit of a snob, an attitude learned early, I suspect. In one of her interviews, talking about her childhood, she said

> The country gentry did not mix socially with the farmers... The children of the gentry could not possibly be sent to the village school... Parents who lived in the great country houses usually preferred to have a resident governess (for their daughters) which was expensive.

The Cambridge University arena was not forbidden territory for her stories, nor were human peccadillos, whatever they were or wherever they may have occurred.

When my wife and I first vacationed in Cambridge, Mary would bicycle down from her garden apartment in Sherlock Close, which was up Castle Hill, just past Fitzwilliam College. She introduced us to tea in the graduate center, where for a few pence one could have tea and biscuits, look out the picture windows, and see the cows in the fields opposite and the punters on the River Cam.

As the years rolled on, she was less inclined to come into town, and we would visit her in her garden apartment. There, her photo albums would come out, stories, scandalous and tame, would get told and retold, and occasionally she would find an old mathematical reprint and explain the gist of it to me. She remembered old friends at Brown and asked me to tell her of their subsequent careers.

On one visit to her apartment, her snobbism came out in a way that I did not initially understand. At that time, I was friendly with Lord Victor Rothschild (1910–1990) (of the famous Rothschild family). Rothschild was a distinguished biochemist, a Fellow of the Royal Society, a man who had had a career at Cambridge before going on to other things and who maintained a house there. (I've described my relationship to Rothschild in *Mathematical Encoun-*

ters of the Second Kind (1997) in some detail.) In the course of my conversation with Mary, I mentioned that my wife and I would be visiting Rothschild and Lady Rothschild later in our trip and asked her if she knew him. She answered along the following lines.

Well, she knew who he was, of course, but she didn't really know him. She had heard that in the years when he was doing laboratory experiments in Cambridge, he was able to short circuit the grants committee's red tape and to provide himself with lab space, equipment and assistance simply by buying them with his own money. It was clear to me that Mary was resentful of this ability. I did not understand the full depth and meaning of her resentment until recently when I happened to read a passage in Chapter V of Aldous Huxley's *Point Counter Point* (1928). One of Huxley's characters is Lord Edward Tantamount, a fabulously wealthy, somewhat reclusive aristocrat, who is also a distinguished biologist. Tantamount conducts experiments in his own home laboratory, and employs Frank Illidge as his assistant. Illidge hates the rich with a profound hatred; hates them more for their virtues than for their sins, and understands his own ambiguous position vis-a-vis Lord Tantamount, a man he admires tremendously.

Illidge was an enthusiastic biologist; but as a class-conscious citizen, he had to admit that pure science, like good taste and boredom, perversity and platonic love, is a product of wealth and leisure. He was not afraid of being logical and deriding even his own idol.

"Money to pay," he repeated. "That's the essential."

It is interesting that as late as the mid-Twemties, pure science was regarded in England as a rich man's game. This, of course, has changed completely.

Now as to what I learned from Mary Cartwright. The year 1968 was a year of turmoil at Brown and at many other American colleges. Among other things, the curriculum was under attack. Brown University was "rescued" from serious disturbances by the curricular proposals of an undergraduate by the name of Ira Magaziner (later intimately connected with the universal medical plan pushed by Hillary Rodham Clinton during her husband's first term as President.)

Mary came to our departmental meetings occasionally, and in the course of one of them, she remarked that when she was a student, all mathematics majors were required to know a proof of the nine-point-circle theorem. Since the nine-point-circle had to that audience the distinct flavor of being beautiful but old-fashioned and irrelevant, I concluded that Professor Cartwright was telling us that we should not be too dogmatic as to what constitutes a proper mathematics curriculum. Fashion is spinach even in mathematics, and time often works to "nine-point-circle-ize" many of our most seemingly relevant and sophisticated topics that are currently insisted on.

Professor Jim Tattersall recently pointed out to me that in the early nineteenth century, the following question appeared on a Cambridge University mathematics exam:

> Give two proofs of the existence of God. Do not use arguments from Revelation.

I like to conjecture that given Mary Cartwright's clerical background, she would probably have done well on that one. Who knows but that if one waits long enough, the question may return to mathematical examinations.

Laughing among the Logarithms:
Adventures of a Speaker

I was once asked whether I considered myself a seasoned speaker. I answered that after one talk I gave I was peppered with questions I was unable to answer. And I allowed my interrogator to take it from there.

Every speaker piles up stories that occurred as a result of the talk or as a result of going to a certain place to give the talk. I've selected two from a whole basketful.

"We have with us this Afternoon"

It was through Emilie's Haynsworth's prodding that I learned something valuable but initially upsetting. There was a member of her department at Auburn University, call him Professor Whitsand, who was a big wheel in the local Kiwanis Club. One day, Emilie told me that Prof. Whitsand had asked her whether she thought I would make an appropriate after-lunch speaker for one of the regular Kiwanis meetings held at the Heart of Dixie Motel.

"He showly would," replied Emilie without having consulted me.

I hit the ceiling when I realized I couldn't get out of this engagement. What ever was I going to say to a pack of Kiwanians? A few standup jokes? No, I would have to say something about mathematics, and, really, if I couldn't find anything to say to a group of intelligent people, how could I justify (to myself) having my mathematical activities supported by society? Well, I thought of something and sketched out a half-hour talk on the back of an envelope.

At 12:00 noon, a group of perhaps thirty Auburn Kiwanians filed into the large function room of the Heart of Dixie Motel. I was escorted in by Prof. Whitsand and was followed by a flagbearer with the emblem of the local club. I was introduced as the speaker of the day by the chairman of the meeting, who added that we had to be out by 1:00 PM. The Chairman then passed out copies of the Kiwanian songbook. Standing, the group sang what I gathered was the Kiwanian anthem and then, with heads slightly bowed, recited a universalistic non-denominational grace. After these preliminaries, we all went over to the hot table where an ample lunch had been laid on. Fried chicken and biscuits, southern style, collard greens, apple pie, and ice cream.

It was now getting close to 12:35 PM. I began to fidget. The dishes pushed aside, we were asked to sing songs numbers 14 and 68 in the songbook. It was now 12:43 PM. My fidgeting intensified. Internally I was getting furious. Externally, I was the pinnacle of placidity.

Songs were followed by the treasurer's report, which *Deo gratia*, was brief. It was now 12:46 PM.

The treasurer's report was followed by the induction of new members, with a short bio read aloud by each inductee's sponsor, followed by the Kiwanian Oath. Two new members brought the clock to 12:54 PM. I had six minutes in which to say something. As a rule, when giving a talk, it takes professors more than six beginning minutes to clear their throats.

I calmed down. I threw away my envelope. I rose and spoke. I told my audience that whether or not they were aware of it, the physical and social worlds were being increasingly mathematized and that the mathematics was largely hidden from view. ("Thank God for that," one of the Kiwanians said under his breath.) I then gave them two examples of significant but hidden mathematizations, and sat down just as the minute hand of the clock moved to the hour.

"I heard your talk at Kiwanis was a great success," Emilie told me the next day.

"It was a good experience for me. Chastening. I'm glad you set it up."

What had I learned? a) Never, never go over your allotted time. Less is more. b) Never, never talk down to an audience. c) It is difficult to explain to a general audience just what constitutes mathematical research but it is necessary to make the attempt, however brief.

Confession in Hobart

Occasionally, one finds oneself in Tasmania. When I speak of "oneself," I am speaking of myself. Naturally. And when I speak of Tasmania, I am not speaking of Tanzania or Romania. I am speaking of the island off the southeast coast of Australia. And when I say occasionally, I mean I have been there twice in a period of eleven years; both times in January, both times in Hobart, the capital of this Australian state, and both times for a mathematical conference.

Hobart in January is without a doubt the most delightful city on the face of the earth. It is summer time. The trees are laden with apples, plums, and apricots. The lavender is in full bloom; the red hot poker plant is incredible. The temperature of the air is moderated by the cold waters of the South Pacific.

The Vice-Chancellor of the University of Tasmania, the institution that was host to our conference, was some sort of a scientist. He was not around to preside at his own cocktail party; off to the Antarctic, looking into the penguins and the currents.

When I got to the V.C.'s cocktail party, people were already working on their second drink. I got mine—an innocent local cider that kicks like a mule after you've had six—and I wondered who or whom I should talk to. There is really no communication problem in Tasmania, as English is the native language of the island, the natives having been killed off some years ago.

I spotted a man at the far corner of the room. He looked vaguely familiar. He was drinking by himself. I went over to him and said, "You look vaguely familiar. Who are you?" (He didn't have his conference tag on his lapel.)

He answered, "I should be familiar. I met you here eleven years ago. My name is William Collins." (I have changed this name for reasons of delicacy.)

"Well, Collins, what cards has the past decade dealt you?"

This was, I thought, a clever metaphor in view of the spectacular gambling casino that had been erected in Hobart in the intervening years. He answered me with a nervous laugh. He went to the bar to refresh his drink and he did not return. A nasty question for anyone to answer. Why had I posed it?

In the week that followed, a week of lectures, and seminars, I kept seeing Collins out of the corner of my eye. At one time I though he was about to come over and say something to me. It could hardly have been about mathematics because our respective fields were too remote from each other. However, I was prepared for a possible encounter. From the conference directory, I learned that Collins was from Adelaide. I would talk to him about the Australian wine industry that centers on this large city. But someone buttonholed me and Collins retreated into outer space.

A week and a half passed. It was near the time for us both to go home— Collins to Adelaide, to the strong sunshine that kisses the grape and turns them into Gewürztraminer, and I to Providence, Rhode Island, another island, so people think, to snow and cold and muck and dark. There was a farewell party. Halfway along, Collins broke through the crowd determinedly and spoke to me.

"You are leaving? Yes? Then I've got to confess something to you. It's been on my mind for eleven years."

"Confess? On this beautiful island? In this lucky land? There can surely be nothing to confess in Hobart."

"You jest. I must tell you."

"Well, I've steeled myself. Make your confession."

"You recall that eleven years ago, we were not staying in this college but in Hytton Hall?"

"I remember."

"You recall that there was a laundry room in Hytton Hall?"

"I do."

"You recall that one afternoon, you came in to do a batch of laundry?"

"I don't remember specifically, but it seems likely. I don't see why I should have left Tasmania with a bag full of dirty clothes."

"You recall that after you did a wash you left and that fifteen minutes later you came back to do another wash?"

"I do not."

"You recall that having come back, you picked up a box of detergent that you had left behind. You found the box empty. You shook your head in confusion and went out?"

"I have no such recollection."

"It's true. The detergent was used up. I stole what was left over."

"You, Collins? You?"

"I was the man."

In the twenty seconds that elapsed between this shattering confession and a necessary response, my kind sought various ways of making that response.

I thought of Dostoyevsky: "Be comforted, Collins. Things are even now. In God's Name they are. Something has been on my mind for eleven years. It was I who seduced your daughter. In the stacks of the mathematics library. Even as she laughed among the logarithms."

"Ah, my Siobhanutschka! Only last week did she turn twenty-three. And a grandmother of two."

I thought of the way of Edgar Allen Poe: "We are all doomed, Collins, doomed. Every eleven years henceforward, we shall return (or our ghosts shall) to this very Conference. And you will make your confession to me periodically. Endlessly. Evermore, or quasi-evermore."

Or the way of Anthony Trollope: "You appropriated, or more properly, misappropriated the residuary cup of detergent? Did you not know that the residue was the bulk of the estate of the Duke of Fabulum, by entail, in the absence of heirs in the male line, destined to revert to the Bishop of Maytag as glebe rent, in perpetuity, and that I was acting only as a surrogate or warden in behalf of my brother-in-law, whose firm handles all detergental matters for the Northumberlandshire prebendaries?"

De Maupassant: "Ha, Collins. Ha. Ha. You could have been spared the agonies of years of silence and repression, years of choked brooding. The detergent you stole was not real. Not real, I tell you. It was only paste. Ha ha, Collins, garden variety laundry paste."

Henry James: "I should have know, Collins, that what with the vertiginous ambiguities and incrustations (not, as might have inconsequentially been deemed superficial) revolving in my mind (as, indeed, one might say) in consequence of its detergentory quality (as it were), the golden bowl of the immaculate, and what is missing from it, has been convoluted and redefined; and the years, the autumns, the winters, the season itself at St. Tropez, would, I should hope, by contrast, be rendered morally ineffectual."

Mickey Spillane: "'You were the rat that copped the cup?' I whipped out the Schweinhundt Silencer that I always take along to conferences. One bullet was left; I had already silenced five lecturers. And I let him have it right through his Tropic of Confusion."

Twenty seconds later, as I said, when I finally found my own voice, none of these words emerged. What came out sounded more like

"Oh, Collins. Collins. *Ego te absolvo.* Go in peace and sin no more. Donate a cup of Rinso to the Girl Guides and send me a reprint when you get back home."

Part VII

Horizons Widen

An Address That Led to Friendship:
Reuben Hersh

You may never know who is in your audience and what his or her reaction has been to what you have said. In January 1964, having won the Chauvenet Prize of the Mathematical Association of America, I spoke at the University of California, Berkeley. The prize was presented at a public ceremony in a large auditorium and I was invited to make an acceptance speech.

What did I say? I've found several old drafts of my remarks, but since I usually wander off from the written version I have in front of me when I speak, I'll have to guess what I actually did say. I know that I didn't say much about the subject of my prize-winning essay whose title was : "Leonhard Euler's Integral: A Historical Profile of the Gamma Function."

Here is a clip from the draft I found:

> The world is being depersonalized. Sociology has replaced bi-ography, and men, lacking identities, wear their names on their coat lapels. Human engineering is taught on every campus, and its essential ingredient, the theory of man-machine combina-tions is expounded with equal status allotted to the machine. Mathematics has contributed its share to this state of affairs and promises to contribute more. It does this through an internal structure that stresses atomized thought (Cartesianism) and the consequences of this atomization touch us from technology to social policy.
>
> The spirit of mankind and the spirit of machines are both fostered by mathematics and it is up to us to balance the valid claims of both.

I wound up making a plea for cultivating both the historical and a socially critical sense in mathematics, neither of which was then or is now a part of a normal mathematics curriculum.

In view of developments in the past three and a half decades, in view of the fact that disembodied voices now speak to us over the phone, in our cars, that biology has introduced novel and depersonalized methods of reproduction, that mathematics, through its applications, has led to social problems that it cannot by itself solve, these words of 1964 seem mild indeed.

They were not at the time. I have the assurance of Reuben Hersh, who was in the audience. Reuben, to whom I was introduced some years later by a mutual friend and with whom I have written two books, told me that he did not believe that at the time anyone would speak to an audience of mathematicians in this way.

Reuben (b. 1927) graduated from Harvard at the young age of nineteen, majoring in English. He won an undergraduate prize in this subject. Some years later, after serving in the armed forces, and after he had worked in a machine shop and in the editorial offices of the *Scientific American*, he became a graduate student in mathematics at the Courant Institute of New York University. When we met, he was professor of mathematics at the University of New Mexico in Albuquerque.

The introduction clicked. From time to time I went to Albuquerque. We drove around; we walked and we talked. I recall our talking at a remote spot on the left bank of the Rio Grande River where there were hot springs. I recall philosophizing with him at the top of Sandia Mountain where hang gliders take off.

We found our philosophical opinions about mathematics generally sympathetic, but far from identical. It was Reuben who introduced me to the writings of Imre Lakatos and Paul Feyerabend, both authors making a great impression on me. Reuben was able to read through a complicated piece of prose and immediately reduce its essence to a sentence or two. We talked about writing a book jointly that would express some of our ideas. Our styles of writing were also far from identical: his style much more blunt and forthright. He had no fear of the colloquial, the gratingly raw or the disturbingly humorous. Mine? Well, the reader can judge it from this book.

After a number of trips, he to Providence and I to Albuquerque, after many, many phone calls, (this was before e-mail was in place), *The Mathematical Experience* emerged in 1981 and has had a considerable success: the right message, apparently, at the right time. The book has been translated into many languages and there is even a special students' edition. The philosophy espoused in this book goes by numerous designations, none of which is entirely appealing to me. Social constructivism is one name by which it is known.

At the time the book came out, the philosophy of social constructivism was quite unpopular; since that time, Reuben has published his own version of this mathematical philosophy, (*What is Mathematics, Really*) and we've been joined by many mathematical philosophers who have arrived independently at more or less the same position. I have in mind, among others, Thomas Tymoczko in the USA; Paul Ernest in Exeter, England; Ole Skovsmose in Copenhagen; Christine Keitel-Kreidt in Berlin; and Glen Pate in Hamburg.

I have no intention of expounding here the principal tenets of my philosophy, or of delineating the individual tenets of other social construtivists. I have written much on the topic and the reader can pick it up elsewhere, to agree, to disagree, to polemicize and, as is often the case with philosophical positions, to misinterpret.

What Are the Dreams?

In February 1978, in the middle of a blizzard that brought the city of Providence to a standstill for more than a week, my wife and I managed somehow to get to the train station and thence to New York. The purpose of the trip was to attend a revival of Massenet's *Thaïs* by the Metropolitan Opera Company, but I took the opportunity to visit the offices of the *Scientific American* and to renew my acquaintance with Gerard Piel and Dennis Flanagan, publisher and editor of that magazine, for whom I had once done some work. In the course of our conversation, I was invited to do an article for them, and I suggested "What are the Dreams of Mathematics?," a topic I had been mulling over for some years.

Mr. Flanagan replied that it sounded pretty good to him and that he supposed that by "dreams" I meant the goals of individual practitioners. I replied that I did not. Nor did I mean nocturnal dreams and their influence upon daylight thought and activity: Kekulé's string of carbon atoms, for example. What I meant by dreams were the timeless, almost mythic drives that have resulted in the creation of characteristic expressions of human intelligence of which mathematics is but one.

Does this sound pompus? Does it sound vacuous? Well, I supposed it ran that risk, but I thought it worth a try.

As I worked up my material, my mind, operating in typical mathematical fashion, began to generalize, and I wondered what the answers might be in other areas. I would write to friends and ask them: what are the dreams of law? What are the dreams of poetry? The important thing would be to distinguish, if at all possible, between dreams and goals.

Goals are rather easily understood. Thus, to cure a certain disease or to find abundant supplies of energy appear as goals of medicine or technology.

The understanding of the basic structure of matter is a goal of physics. But when a mathematician sits down to crack a problem, he doesn't often think he is creating a universal language; nor does a lawyer, when he handles a case, have the image of justice constantly in front of him. Nonetheless, the absence of these two tendencies would put both the mathematician and the lawyer out of business immediately.

I received answers in abundance. Among the more interesting responses was the one from George Wald, who received a Nobel Prize in Physiology in 1967 for his work on vision. Here are several clips from his letter.

May 17, 1978

Dear Phil:

You ask for the dreams of biology....

What we need to understand, and most evade in our preoccupation with reduction, fragmentation and analysis, is the wholeness of living organisms and of their societies and environments...

I have worked all my scientific life on the mechanisms of vision. The field is now exploding, and we have learned and are learning a lot. I want to say this: one can put all of that together and extrapolate it endlessly assuming complete successs in all our present efforts, and it says nothing of what it means <u>to see</u>. That is as though in another universe that we know no way to approach. Yet embarassingly central: all that we know and learn, all our science, is rooted in that consciousness of which we know nothing. We know it alone, yet nothing <u>about</u> it. That is, of course, where mathematics happens too. Could one figure as well if one didn't <u>know</u> that one was figuring? Like a computer? Does a computer know? How could one find that out?

Thanks for making me dream.

As ever,

George Wald.

(Since this letter was written, many attempts have been made to explain consciousness based on a mixture of physics, computer science, mathematics, physiology, and psychology. I will allude to Sir Roger Penrose's work in this area later in this book.)

I also wanted to obtain the thoughts of a theologian, and for this purpose, I could think of no better, no more accessible person to ask than my father-in-law, Louis Finkelstein. Here, very slightly abridged, is the letter he sent me in response to my request.

May 9, 1978

Dear Phil:

I was interested in your comment that most mathematicians who are philosophers are also Platonists. I myself noticed this many years ago. Being myself deeply influenced by Morris R. Cohen, I also tend to think of Universals and Relations, as well as Ideas, as more real than their reflections in this shadowy world. However, I have not found many other people who would agree with this.

Now, as for the dreams of this particular religionist. They are really questions that I often ask myself, although in my youth, I used to put the issues more positively, as things which I hoped myself to help realize in this tough world.

1. Can we ever develop a civilization in which everyone, or almost everyone, will be devout in his own religion, and totally committed to its theology, and yet tolerant and understanding of others?

2. Can we develop a world or a civilization in which to ask this question is not considered heretical?

3. Can we develop a civilization in which the most pious people are also the most moral? It ought to be axiomatic that a religious person behaves properly and lovingly towards his fellow man.

With much love,

Louis

In the years that have passed, I have thought of this letter often. I tend to reformulate it by saying it is the dream that the pious should be good and the good should be pious. But this is by no means correct or adequate, for logically it might be fulfilled by the existence of a small subclass of humanity that was simultaneously good and pious while the rest of humanity went hang. Louis Finkelstein surely believed in universals not only in the philosophical sense, but as a total human aspiration. The legend of the Thirty-Six Righteous People, a tiny subclass of the righteous and the pious, for whose sake God maintains the world, would have been for LF a myth whose value lay in its pointing to a universal possibility.

In the next section I will reproduce the reply of Sir Alfred Ayer, a distinguished philosopher, but here is my own 1978 answer for mathematics after I asked myself the question.

1. The Dream of the Universal Language.

Mathematics is the universal language, independent of culture, and is applicable in all intellectual discussions.

2. The Dream of Indubitability

The statements of mathematics are absolutely and universally true.

3. The Dream of Infinity

Mathematics has created infinities of many sorts. It manipulates them, uses them, and makes them palpable.

4. The Dream of Description and Oracularity

Mathematics, as an adjunct to other disciplines such as astronomy, physics, biology, or economics, can describe the present and predict the future.

These dreams—they have also been called myths—represent some of the most powerfully energizing intellectual principles known to the life of the mind.

My projected book *What Are the Dreams?* was never published. It was never written, though I received many replies and had even selected for epigraphs two quotations that reflect the passage of two centuries:

> When I was young, the *Summum Bonum* in Massachusetts was to
> be worth ten thousand pounds Sterling, ride in a Chariot, be
> Colonel of a Regiment of Militia, and hold a seat in His Majesty's

Council. No Man's imagination aspired to anything higher beneath the Skies.
    –John Adams to Thomas Jefferson.
    Letter of November 15, 1813.

The dream of vulnerability conquered is the dream of the age.
    –Josephine Hardin, *The Vulnerable People*, 1978.

Freddie Ayer

The Providence Athenaeum is an old-line private library just exiting from the throes of computerization that half of its clientele would preferred not to have happened. The library is full of ghosts, the most prominent of them being Edgar Allan Poe who read books there while fruitlessly pursuing Helen Whitman who lived up the street.

I first met Valerie Hayden at an Athenaeum party when Hadassah and I shared a table with her and her psychologist husband Brian. It was immediately apparent to us that Valerie was British, and was an exotic flower among the beds of petunias that populate the Providence gardens of personalities. Since she was young, bright, sharp, beautiful, charming, and established an immediate presence, I'm sure that men fell in love with her right and left. I, personally, could not see enough of her, but actually we were able to get together only a few times a year. She was very busy as curator of prints at the Rhode Island School of Design Museum.

She was born Valerie Ayer, and was the daughter of Sir Alfred Ayer, a British philosopher of great reputation, and a TV personality. Through Valerie and her husband I established a bit of a connection with her father via letters and phone calls.

Freddie Ayer (1910–1989) (for so he was called familiarly) had been a *wunderkind*. At the age of twenty-six, with the publication of his book *Language, Truth and Logic*, which burst on the British philosophical scene with great eclat, he established a solid reputation as a positivist philosopher. An Oxford don for many years, suffering a bit from his precocity, Ayer must ultimately have got bored with pursuing academic philosophy, for he joined the "Brains Trust," a information-type radio show, where he was able to swap wisdom, knowledge, and wise cracks with other high-table wits. Besides philosophical works, he wrote an amusing autobiography in two volumes.

I was familiar with *Language, Truth and Logic* and though realizing the great splash it had made in 1936, I found it too positivistic for my tastes. What I wanted from Ayer in my correspondence with him was not a reiteration of his published philosophy, but a contribution to my projected collection *What are the Dreams?* Ayer very kindly obliged in a letter a part of which I have often quoted.

New College, Oxford
9 June, 1978

Dear Professor Davis,

You ask me for a list of philosophical dreams. Different philosophers would give you different answers, but I suggest the following:

1. To find a secure foundation for our claims to knowledge.

2. To agree upon a criterion for deciding what there is.

3. To draw a distinction between necessary and contingent truths. (The truths of mathematics have commonly been taken to be necessary, but this has been disputed.)

4. To reconcile the scientific account of the physical world with the evidence of the senses.

5. To achieve an analysis of judgements of value.

6. To develop an adequate theory of signs.

Yours sincerely,

(signed) A. J. Ay (sic)

Sir Alfred Ayer
Wykeham Professor of Logic

The parts of Ayer's letter that I latched onto were Numbers 1 and 2, and I usually phrase them as: "What *is* there, and how do we know it?"

I repeat: the question of what is and what isn't nonsense is not so easily resolved. What contributes to the difficulty is the possible historic transformation of sense into nonsense and of nonsense into sense. To many in the sixteenth and seventeenth centuries and before, the evidence for the existence of witches was clear and incontrovertible. To us, the idea is nonsense.

And it proceeds in the other direction also. The earth goes around the sun? Why, what nonsense! Gian-Carlo Rota, in a lecture on the phenomenologist philosophy of Edmund Husserl, pointed out that Husserl believed that if "new rigorous theoretical sciences come into being" (say, in the social areas) by new Galilean Revolutions, they will be characterized by "(a) The denial of common sense and (b) theoretical laws."

Comes the Revolution, then, nonsense metamorphoses into sense!

(I can't resist interpolating at this point that Ayer's question occasionally comes up in contexts other than those of ivied philosophy buildings. In 1998, President Bill Clinton, about to be impeached for lying under oath, raised the question "What is 'is'?" Every undergraduate in the country knew what he was talking about, but the pundits and the smart-alecks made a field-day with it.)

In those years, my hobnobbing with philosophers led me to receive an invitation to lecture to the Oxford Philosophical Society. I thought of all the Oxonian philosophers who might attend my lecture or whose ghosts might be present—Gilbert Ryle had died within the decade—and I, who consider myself at best to be a kibitzer in philosophical fields, began to have cold feet. And then I thought of something that led me to accept the invitation: the words of a rank outsider do not "count." I could get up in front of these people, make a fool of myself if they happened to think so, and the locals would simply write it off as a bad investment of their time.

I had noticed this years ago when first visiting England where a fairly strong caste system is still at work. By declaring myself as an American and hence as an outcast who didn't "count," I was released from the assumptions, presumptions, and self-limitations of the natives. I could even invite a bartender of my acquaintance to mix socially with a St. John's College, Cambridge graduate and get away with it. (But the tension was heavy, I'll tell you!).

In my talk to the Philosophical Society, I discussed an aspect of my version of the social constructivist philosophy of mathematics. In the course of my talk, I probably said that I saw no necessity and no profit in distinguishing (as Ayer had suggested in his point No. 3) between necessary and contingent truths.

The talk went well, with about fifteen in the audience. Then I learned something about philosophers that I hadn't realized: real philosphers love to argue. Public argument is the breath of life to them, their mother's milk. They have a tradition for this going back to the Peripatetics, if not earlier. I could sense that my audience were all waiting on tenterhooks till the moment I shut up, the moment when they could then rip mercilessly into what I had said.

I, on the contrary, tend to avoid public argument. I want to make my points on paper or in a talk, and will answer attacks in a paper but not during the question-and-answer period. As soon as I saw that all were prepared to put me (and each other, for that matter) on the rack, foregoing their supper if necessary, I bowed out of the procedure by hiding behind my status as an outsider. The lecture started at 4:30 P.M. and I was back with my wife by 6:15 P.M.

It was not so very long, it seemed to me, after our first meeting that I received a long-distance call from Brian Hayden. I was then teaching a summer course on the philosophy of mathematics at Wesleyan College, in Middletown, Connecticut. Would we come to a memorial service for Valerie, to be held in a few days in the Manning Chapel at Brown?

What? Valerie gone? How can that be? It was only a few... Of course we'll come. The poem she loved was sung: Blake's "In England's Green and Pleasant Land." One never knows fully what one has meant or what one has contributed to the life of another person.

Consider the Body *B*

Mathematics is not, contrary to popular opinion, all numbers and formulas. These are not the only mathematical nouns. There are figures (e.g., planes, triangles, polyhedra), there are structures, algebraic and otherwise (rings, groups, fields, function spaces). There are adjectives that qualify (infinite, commutative, complex); there are verbs (add, multiply, differentiate, construct, satisfy). A good mathematical dictionary would contain more than five hundred pages of specialized terms, a very good one would go to fifteen hundred.

Mathematical language has its own syntax, semantics and semiotics. Though absolutely crucial in a certain sense, the nouns, the verbs etc., are the least of it; what matters are the ideas that lie behind these words. Natural languages such as English, French, Danish, etc., play a great role in mathematical texts in setting forth and explicating the ideas. A page of raw, naked, symbolic mathematics by itself would be absolutely uninterpretable.

Consider an n-dimensional Euclidean space. Consider a unique factorization domain D. Consider the automorphism group gamma of a graph G. Consider a Hausdorff space. One of the very frequent natural language expressions found in mathematical expositions is the verb "consider."

The word "consider" in such contexts has a variety of uses. It serves to pin down a notation. It serves to bring up a *mise-en-scene* very much as in a dramatic production. It asks the reader to bring into his or her immediate memory all that he or she knows about a "unique factorization domain" so that the reader does not go into shock when the writer pulls out one or two special facts and uses them in further argumentation. The use of the word "consider" in mathematics has been the object of deep semiotic contemplation. Thus, semiotician Brian Rotman has written

The imperative "consider a Hausdorff space" is an injunction to establish a shared domain of Hausdorff spaces; it commands its recipient to introduce a standard, mutually agreed upon ensemble of signs—symbolized notions, definitions, proof, and particular cases that bring into play the ideas of topological neighborhood, limit point, a separability condition—in such a way as to determine what it means to dwell in the world of such spaces.

Consider a Hausdorff space means view attentively, survey, examine, reflect, etc.; the visual imagery here being part of a wider pattern of cognitive body metaphors such as understand, comprehend, defend, grasp, or get the feel of an idea or thesis.... To speak of dwelling in a world of Hausdorff spaces is metaphorically to equate mathematical thinking with physical exploration. Clearly, such worlds are imagined, and actions that take place within these worlds are imagined actions.

With these words, Rotman comes close to the personal psychology and inner imagery of doing mathematics.

One of the instances of the injunction "consider" that has always amused me appears in aero- and hydrodynamics as well as in the theory of elasticity and is: "Consider the body B."

Now what is a body—a very suggestive choice of a word—mathematically speaking? Without going into deep topological definitions, we may give examples of, say, three-dimensional bodies: a cube, a pyramid, a sphere, a doughnut, a pretzel, the surface and interior of an airplane or an automobile. The letter B, of course, stands for body (it may have originated with the German "Bereich," meaning domain) and is often used to designate a specific body you have called to mind from out of a wide range of possibilities.

Considering the body B in the context of continuum mechanics means that you know or you assume a variety of things about it; that it has a center of gravity and a moment of inertia, that you can erect perpendiculars to its surface at most of its surface points, that you can apply Green's and Stokes's theorems to it, etc. Sometimes, when authors want to illustrate a body B in their writings, they draw a Dali-esque or a Hans Arp type of figure to stress the generality or freedom of the concept that has loosed the bonds of geometric rigidity seen in cubes or spheres.

I have been so amused by the phrase that some years ago, when my colleague Prof. Harry Kolsky, author of *Stress Waves in Solids*, was ill in the

"The Body B," a painting by Phil Davis.

hospital, I painted a picture in the Renaissance style, depicting the apotheosis of the body *B*, and gave it to him for his entertainment.

One January of the early Eighties, I was invited to a conference on numerical analysis held at the University of Dundee in Scotland. Though the days were short, roses and other flowers were in bloom and the heavy winter clothes I had brought along were unnecessary. I had not been informed of the Gulf Stream. I arrived on January 24th and my host immediately informed me that on the evening of the 25th, a birthday party for Bobby Burns would be held. I was more than welcome to come providing I had some decent clothes. Would a dark suit do, I asked. It would, so I went to the party.

As I recall, the party was held in one of the university facilities with an attendance of about twenty-five, mostly mathematicians and their spouses. Two long tables had been set up. Some of the men and women were dressed in kilts and bonnets, sporrans, and had dirks stuck into the tops of their calf-length argyle socks.

The meal and the whole proceedings were quite ritualized, so I was informed, and we started off with cock-a-leekie, a thin broth made from chicken and leeks. This was followed by a fish course: cold poached salmon. These two courses were mere appetizers that served to introduce the *pièce-de-résistance*.

A procession formed in an anteroom and made its way to where we were sitting. It was headed by a bagpiper in full Scots' dress, of course. The piper was a mathematician borrowed from St. Andrew's University just across the Tay Estuary and was followed by a man in a chef's hat, holding aloft a large silver tray on which sat a large amorphous-looking brown object that vibrated very slightly as it was carried in. The brown object was followed by two men, also in chef's hats, each bearing aloft two bottles of usquebaugh (whiskey).

"Make way for the haggis. Make way for the haggis," cried the whiskey bearers above the piper's notes and the procession snaked its way around the tables and came to rest at a central position where the master of ceremonies sat.

The haggis, for such is the designation of the brown object, was set on the table and the master of ceremonies poured out a double whisky, one for himself and one for the piper, which they downed in one gulp. Then the master of ceremonies raised his dirk high. At this point, the crux of the ritual, the company stood up. The MC recited from memory and in a most dramatic manner Robert Burns's "Address to the Haggis," after which he plunged his dirk into its heart. A slight squirt ensued, followed by a rush of steam, and savory, sensual odors (to some, anyway) diffused among the diners.

The whiskey bottles were passed around and more were fetched from a case. After the Selkirk Grace was said, and the haggis was cut up and distributed accompanied by ample quantities of clapshot (natties, i.e., mashed potatoes mixed with neeps, i.e., mashed turnips).

While the main course was being eaten, individuals rose and recited their favorite Burns poems, or told stories of a quite arbitrary nature. The native Scots present referred generously to the presence of numerous Sassenachs (= Saxons = British = foreigners including Americans) and explained to them the significance of the ritual.

As an invited guest from overseas, I was asked to contribute some remarks. I demurred, but the company insisted. What on earth should I talk about, I asked myself. I knew no Burns by heart. An idea occurred to me. As the haggis was marched in, resembling a huge brown lima bean suffering from elephantiasis, it reminded me of the many "Bodies B" I had seen illustrated in texts on continuum mechanics. This was my cue and I got up and said something along the following lines.

> The letter B is for Burns and the letter B is also the frequent designation for three-dimensional objects in advanced texts on continuum mechanics. The object we have been consuming is a Body B [Ooh! Ooh! Hear him now ! Not an object!].

Now consider the Body B. It is neither an perfectly elastic solid nor a fluid. Nor is it a plastic object. It is a unique thing whose very uniqueness defies traditional mathematical analyses. [Hear, hear! Ray!]

It flows, transforming itself; it vibrates. It has time-dependent eigenvalues (important parameters in vibratory analysis). But being constructed as it is, of oatmeal, ground liver and other strange glands, discarded argyle socks, and fat, bound together by the bladder of a sheep [No! Stomach, not bladder!] stomach, then, and cooked till mild solidification takes place [Hear, Hear!], I set as a challenge for the numerical analysts present at this Burns celebration, the problem of the accurate and expeditious computation of the eigenvalues of this Body B.

The whiskey glasses were refilled and raised. A cheer went up. "To the eigenvalues. To Robbie Burns and the Body B."

I sat down. I lasted until 12:30 AM, and then begged off, pleading jet lag, which was no joke, and returned to the dormitory room assigned to me by the conference.

Taking Instruction with Gian-Carlo

One day after listening to me hold forth, Gian-Carlo Rota, combinatorialist extraordinaire at MIT, philosopher, belletrist, and gourmet told me that I was a closet Husserlian. Why didn't I admit it and come out into the open? I hardly knew what he was talking about, for I knew the name of Edmund Husserl (1859–1938), mathematician and philosopher, an intellectual parent of Heidegger, only as one that I might have seen out of the corner of my eye while flipping through the *H* volume of an encyclopaedia.

I told Rota that I was glad that somebody had pinned down what my philosophy was, for I was always at a loss when asked about it.

"Husserl is the greatest philosopher since Kant. So said Gödel. If you want to know what Husserl is all about why don't I tell you? Come up to Cambridge, we'll have lunch, and I'll tell you."

And that's the way it was. At intervals of several months over a period of several years, I would go to Cambridge, we would have lunch at MIT or across the river on Charles Street in Boston, where excellent gourmet lunches were to be found (e.g., strange sea things garnished with strange land things), and Gian-Carlo, one of the leading students of Husserl in the world, would tell me about his favorite philosopher.

We talked back and forth, and when the going got tough, which was soon enough, I found I didn't understand a word. Rota, I thought, was giving me the cold turkey treatment. It was like being plunged into the midst of one of the more esoteric tracts of the Kabbalah, not understanding the terms, let alone the original language, not understanding what the philosophic or metaphysical goals were, not understanding the manner in which these goals had or had not been attained.

Gian-Carlo Rota.

In the course of "taking instruction" from Rota, I met his friend Bob Sokolowski, priest and professor of philosophy at Catholic University in Washington. I had a lovely meal and talk with Bob on M Street, Georgetown, and understood very little of his brand of phenomenology. From Sokolowski I heard that Pope John Paul II, when he was simply Karol Wojtila, was a phenomenologist and had contributed articles to the *Journal of Husserl Studies*. I looked one up and was baffled. As regards reading Husserl in the original—translated into English, of course—it was to quote Sam Goldwyn: "I'll tell you in two words: IM POSSIBLE."

At one point during these months of instruction, I said to Gian-Carlo: "I'm not getting it. I'm simply not getting it." And Rota answered me, quite gently, not disappointed at all. "Give it time, give it time. One day, you'll be walking along the street and suddenly you'll say to yourself: "I've got it!" The "Aha" phenomenon.

My answer to that was a hasty "It's all right for you, G-C., if you have a personal illumination. But as a teacher, you have to communicate. That's what language is supposed to be all about. If you can't communicate, it's as though your words, written or spoken, never existed." Privately, I said to myself that whenever ideas are put forward, one has to come to one's own understanding of them through whatever means. But that day may never come around.

I think that Rota took my admonition to heart, for over the following years, he has made concessions to aid the untutored in philosophy. I am only now beginning to understand not Husserl but Husserl-Rota and I admit to being a semi or a partial or a weak Husserlian. Insofar as I can understand him, I find Husserl-Rota too much of a Platonist for my taste.

Most recently Rota gave a lecture on Husserl and mathematics that included a list that in some manner approached the ideas in my "Dreams of Mathematics." Interpreting Husserl for us, Rota asks what is it that characterizes mathematics? It is:

(1) absolute truth
(2) items, not objects
(3) non-existence
(4) identity
(5) placelessness
(6) novelty
(7) rigor.

I pass this list on to the reader but will not attempt to explicate the several items (not objects!), as Rota did in his lecture. To tell the truth, I find these items much easier to understand than original Husserlian concepts such as *Fundierung* or the *primordial Nicht*.

In the fall of 1998, I read in the papers that Pope John Paul II had just issued an encylical *Fides et Ratio* (*Faith and Reason*). I wrote to Bob Sokolowski for the English version, and through Bob's courtesy and that of the Web, I received it rapidly.

What I had in mind by my request was this: since mathematics was regarded by most of the world as the pinnacle of reason, and since I believed that a large component of faith was also present, perhaps I might be able to read the words of John Paul, a Husserlian, as coming close to such an assertion.

In reading *Fides et Ratio* I was immediately thrown into a world of discourse that was largely unfamiliar to me. But I got this much out of it: unreasoning fideism as well as secular rationalism are rejected (the Church's traditional position). Thomas Aquinas is demoted from his nineteenth-century status as *the* philosopher. In fact, the whole Western philosophical position is relativized, suggesting that it is one tradition among several (e.g. the Chinese).

It also makes some moves towards Kant (e.g., it asserts absolute moral duties). With my imperfect understanding of phenomenalism, I could not detect any whiff of it in the encyclical. Nonetheless, the mere juxtaposition of faith and reason in a public pronouncement made me consider once again what their relationship might be in the philosophy of mathematics.

Continuing Education:
Isaiah Berlin and Giambattista Vico

It was in the mid-Fifties that I first heard of Isaiah Berlin (1909–1998) as a great thinker and writer, but it was not until a quarter of a century later that I'd read anything he'd written. When I did, I found that my ideas, where they overlapped, were in such considerable consonance with his, that I felt myself as a younger but totally uneducated *doppelgänger*. How did this agreement come about? Was it the *Zeitgeist* at work, or was it something much less mysterious?

Berlin (later Sir Isaiah), Fellow of All Souls, Oxford, philosopher and brilliant historian of ideas, was one of the most learned and deeply humane men I have ever met. I like to tell people that of all the academics I've known, his head was the one that was really screwed on right. The media made much of Sir Isaiah at the end of his life: he loved to play the guru. But no matter.

Reading Berlin led me, through his long essay on the man, to Giambattista Vico (1668–1744) a Neapolitan lawyer and philosopher. (*Vico and Herder: Two Studies in the History of Ideas*.) Just as Felix Mendelssohn had reclaimed Bach, Berlin was one of the scholars who reclaimed Vico for our generation. Vico himself is enormously difficult reading—forbidding, even—and what I know of his ideas came to me through Berlin's interpretation.

When I read the following passage from Berlin's Essay, it was to me like the smell of coffee in the morning.

> (Vico) became convinced that the notion of timeless truths, perfect and incorrigible, clothed in universally intelligent symbols, which anyone, at any time, in any circumstances, might be fortunate enough to perceive in an instantaneous flash of illumination, was (with the exception of the truths of divine revela-

tion) a chimera. Against this dogma of Rationalism, he held that the validity of all true knowledge, even that of mathematics or logic, can be shown to be such only by understanding how it comes about, i.e., its generic or historic development. In order to demonstrate this, he attacked the claims of the Cartesian school in the very field in which it felt itself strongest and most impregnable.

Vico conceived of the idea of the variety of human cultures whose thoughts, languages, practices, aspirations, were essentially non-translatable from one to the other. According to Berlin, he was the first philosopher to have done so. My own meandering thoughts had led me some years before reading this passage to similar conclusions with regard to the nature of mathematical knowledge, conclusions that were anti-platonic, anti-Cartesian, and I felt that my own stance was bolstered by Vico's insights and what I took to be Berlin's tacit approval of them.

The platonic philosophy of mathematics did not square with my own experience either of learning mathematics, creating new mathematics or appreciating the history of the subject, and I had worked out in my mind, though never set down in systematic fashion, a theory of mathematical evidence that embraced experience, logical deduction from hypotheses, cultural influences, historical change as part of a larger whole. By the early Sixties I was ready to get up and say in public that the history of the subject, rarely taught, rarely discussed, was an important part of one's understanding of mathematics.

In May 1990, I was able to meet Sir Isaiah in person. (I've described this meeting in some detail when I wrote of my relationship with Lord Victor Rothschild in *Mathematical Encounters of the Second Kind*). I told Berlin how much I had learned from his books. He smiled. (Praise me, I can stand it!) I went on to expatiate on my own Vicovian theory of mathematical knowledge that took off from the paragraph by Berlin that I have just quoted. Berlin, a rapid, non-stop talker, but not a conversation hog, remained silent for a few moments nodding his approval all the while.

I take it that there was little more Berlin wanted to add to that particular subject, for when I had stopped, he burst into a recitation of all the mathematical luminaries in Europe that he had known personally. It was then easy enough to establish personal connections with numerous third parties (as one usually does when meeting for the first time). I walked away from my appointment happy. I had had my Warholian moment in the sun.

Giambattista Vico.

Perhaps the pinnacle of my Vicovian career in England came in January, 1998, when to celebrate the Fiftieth Anniversary of the death of Alfred North Whitehead, the department of mathematics at Imperial College, London, where Whitehead had been from 1914 to 1924, asked me to deliver the First White-head Memorial Lecture. Though I did not mention his name explicitly, I felt that the Ghost of Giambattista Vico was sitting in the back row and nodding his head in approval before my droning lulled him back to sleep.

Part VIII

Copenhagen and Its Aftermath

Bungling In Budapest:
A Second Address That Led to Friendship

The International Congress for Mathematical Education, (ICME) meets every four years in different cities of the world. In the summer of 1988 it met in Budapest. The invitation committee asked me whether I would give a half-hour talk and I was agreeable. Hungary had just been opened up politically and to Western ideas, the free market, etc., and my wife and I thought it would make a particularly interesting trip. I planned to talk on *"Applied Mathematics as Social Contract"* and received notice that I was scheduled to go on at 10:00 AM, Tuesday, in such and such a room.

We were put up in a fairly new dormitory building owned and operated by the Hungarian Computer Society. A bit cramped, but adequate. At least I had room to take off my coat without going into the hallway, which is more than I can say for some of the hotels I've stayed in recently in the Kensington section of London. The ICME sessions were held in the Technical University on the bank of the Danube, a nineteenth-century building sadly in need of upgrading. I recall passing the office of what in the USA would be called the university president and seeing the sign "Rector Magnificus" over the door. How wonderful, how medieval, I thought.

On the Tuesday, I arose early, breakfasted downstairs in the dining room, looked over my lecture notes, and took the tram to the university. I arrived there early. I selected a lecture room at random from one of the numerous parallel sessions. I sat down and listened to the speaker. Perhaps it was in German, perhaps in French. I fell asleep. Blame it on jet lag and not the droning speaker; jet lag is a most convenient and legitimate excuse. The next thing I knew, I was being shaken by Reuben Hersh.

"Where've you been? I've been looking all around for you. You know you were to supposed to speak at 10 in Room 308?"

"My Gawd! What time is it now?"

"Ten after."

Reuben pulled me by the ear (so to speak) down to Room 308.

At conferences, talks are scheduled very tightly. God forbid that any speaker should run over. Fifteen minutes of my allotted thirty were already shot. But I lucked out. Despite—or perhaps because of— my memory lapse, my brain cells got their act together, and I presented the heart of my paper in half the time. I said, in effect, that *the time had come for mathematics to stop pretending it had nothing to do with people*. One way that I often put it now, parallelling the epigraph to this book by Edith Wharton, is that mathematics is a species of fiction, created by humans, and is not morally neutral.

My abridged and energized talk was a great success: less is more. People came up and congratulated me. My tardiness was forgiven and forgotten. Numerous further invitations derived from my bungling in Budapest. One came from Denmark, and I am about to describe its sequellae, one of which is the present book.

The Copenhagen Connection

In the summer of 1990, I was invited to give an address at a Danish conference on mathematical education. I suppose my book with Reuben Hersh, *The Mathematical Experience*, together with my Budapest performance had impressed someone on the invitation committee. The conference was to be held at Roskilde University, about twenty miles west of Copenhagen.

Roskilde is an ancient city at the tip of a fjord, and in centuries gone by was the capital of the country. Some of the old kings and queens of Denmark are buried in the Roskilde Cathedral and Viking ships lie buried in the sands of the fjord—some of them now dug out, reconstructed, and displayed. As regards the university, when I accepted the invitation I did not know that a few years before, it had been in the forefront of social activism and student protest, so much so that some administrators and politicians in the Danish government proposed to close it down.

This was the first time my wife and I had been to Denmark, and the conference put us up in the modern dormitory of the Roskilde Slagteri Skole, a nearby trade school for the Danish meat industry. (Denmark exports a considerable quantity of meat products.) Of course, in writing to family and friends back home, it was easy to make the obvious jokes; how all schools butcher their students..., etc.

I've not kept my notes for that talk. What I most likely said was that mathematical education must be relevant to the day-to-day concerns of students. But more than this, it ought to be aware and to reflect critically on the social implications of the increasing number of mathematizations that had been installed in the past century and were still being installed daily.

The first point was hardly new. Alfred North Whitehead made it around 1910, but the lesson, despite computerization, has not yet been entirely learned.

Changing a curriculum, it would seem, is a harder job than relocating a grave-yard. Higher education in all countries is an absolute stronghold of self-protec-tive conservatism.

My second point, the one I made in Budapest augmented by the plea that mathematics curricula at all levels should inculcate an awareness of how math-ematics alters society, had already been taken up by some of the Danish educa-tion theorists, and they were quite pleased to have their thoughts reverberate in the independent words of someone from across the Atlantic. In short, my talk went well. So much so, that I received a note that a Professor Boos-Bavnbek, who had been in the audience, wanted to have breakfast with me the next day to talk over some ideas he had. Agreed. I invited him and his wife to join us in our dormitory breakfast room.

At 8:15 AM, we found Bernhelm Booss-Bavnbek and his wife Sussi downstairs waiting for us. Cheerful introductions and a warm welcome from them to Roskilde. The Slagteri Skole laid out the usual splendid morning spread, typical of central Europe, but with a special Danish flavor. Juices, rolls and breads of many types, cheeses, twenty kinds of cold cuts, soft boiled eggs hidden cunningly underneath a large hen-shaped egg cozy, coffee, fruit. What more does the stomach require at 8:45 AM to send a proper message to the brain?

Before going on with my story, I should say a few words about Bernhelm— I use his first name because we subsequently became friends. He was born in Germany in 1941 to a family of lawyers. He received his degree of *Doctor Rerum Naturae* in 1971. Early on, perhaps in revulsion against the events in Germany in 1923–1945, he became a devoted Marxist. A fiery speaker, and a fiery author and pamphleteer (he is the closest personality I've known to our own Tom Paine of *Common Sense* fame), he took part in the student uprisings of the late Sixties and Seventies. He had a university position in Germany but his activities resulted in a *Berufsverbot* (a cancellation of professional job opportunities) which left him unable to pursue his mathematical career in his native land.

At that time, nearby Denmark was more open politically than Western Germany. Bernhelm moved to Denmark and found a job in the newly founded Roskilde University Center. A devoted and a deep mathematician, he pursued simultaneously the theory of partial differential equations, particularly those parts related to the work of Sir Michael Attiyah in England, as well as a full life of social agitation. He rceived the Boerge-Jessen Award of the Danish Math-ematical Society in 1980.

Mogens Niss and Bernhelm Booss-Bavnbek, Roskilde, Denmark, 1997.

Political (but not ethical) Marxism was a dead duck by the middle of the Eighties, and Bernhelm turned to Green concerns. The blessings of technology were great. The sins of technology were great. How to steer one's life to a compromise? The ways in which Bernhelm conducted his personal life, I thought, were often draconic. For example, he didn't own a car and he wouldn't eat canned food. By way of a tease, I once brought him and his wife Sussi a can of peaches that I bought at the convenience store at the local S-Tog station, and I'm sure the can is still sitting in their kitchen cabinet. This green super-orthodoxy in no way interfered with our friendship.

Bernhelm had the idea that he and I might collaborate on a book that would elaborate our apparently coincident views on the role of mathematics in society. We would raise critical questions. The plan seemed to me to be both reasonable and important, and we agreed to be in touch to work out how we might get together over a longer period and develop the contents of the book.

In the course of the breakfast, Bernhelm told us about his travels in the US, and how the person of Thomas Jefferson had caught his imagination. Our book, Bernhelm thought, might take the form of a message addressed to the "Ghost of Jefferson," in which Jefferson's opinions would be discussed against the background of what has happened since his time. Going beyond, we imagined that we would tell Jefferson's Ghost how the whole of mathematics had developed since his day, and what its relationship to society now was.

It was Bernhelm's idea to feature Jefferson as an Enlightenment man who believed that mathematics exerted a moral force, and to feature Rousseau as one who believed the opposite, that science corrupted morals. Though the Rousseauvian, Thoreauvian, Luddite concerns with regard to science and technology had been set forth by many contemporary authors, taking the argument down to mathematics, the very basic language of theoretical science, had been largely neglected. The proposed book might very well have the title *Updating Mr. Jefferson.*

After I met Bernhelm, my own admiration for Jefferson joined his. Jefferson was one of my favorite characters—what American does not see him as a hero and an icon?—and I was aware that he had probably studied more mathematics than any president who had sat in the White House (with the possible exception of Jimmy Carter who was a U.S. Naval Academy graduate). I knew also that the 250th Anniversary of Jefferson's birth was coming up—it is exactly halfway between Columbus's discovery of America and the present. Bernhelm's plan therefore seemed agreeable.

After my return home to Brown University, our friendship and collaboration was enhanced and facilitated by fax and e-mail. It's too bad that one can't as yet enjoy a real breakfast with one's far-off friends by fax. Perhaps that will come.

Nyhavn 18

In the spring of 1992, Bernhelm arranged for me to present a number of lectures on the relationship between mathematics and society at the University of Roskilde in Denmark. Most unusual accommodations in downtown Copenhagen were provided, which I have described in the story *Thomas Gray in Copenhagen* (1995). I quote from my book:

> Nyhavn 18 is an eighteenth century building situated centrally in a most picturesque part of Copenhagen, just off the harbor. In 1968, it was bought by Denmark's National Bank (which corresponds to the Bank of England or the Federal Reserve Bank in the US), and renovated for the purpose of providing flats for foreign scholars, artists, scientists, and writers for a period of up to a year.
>
> Hadassah and I were fortunate enough to be accommodated in one of these flats. Our pleasure was further enhanced when we learned that from 1873 until his death in 1875, Hans Christian Andersen, the famous writer of fairy tales, had occupied two of the rooms of our flat. This information is widely available. Tourists can find the address in the useful *Copenhagen This Week* pamphlet given out free of charge at every hotel. The bank had underlined this information for occupants of the flat by assembling and placing on its walls some Andersen memorabilia. Although we were living in a mini-museum, it was not open to the general public.
>
> As I sat at a beautifully modern, carefully crafted desk of Danish design, I would look out the window to the other side of

the canal and see the masts of the sailboats docked there. If I looked up at the wall in front of my desk, I would see a photograph of Hans Christian Andersen sitting in the same position at his desk (of heavy mid-nineteenth-century design) looking out the same window at substantially the same harbor scene.

Every morning I would ride the S-Tog (the local train) to Roskilde, a half hour to the west. In the late afternoon I would return to Nyhavn 18. In the evening, I would straighten out my lecture notes for the next day and then commune as best I could with the ghosts of Hans Christian Andersen and Thomas Jefferson, two totally different kinds of people.

Part IX

Interlude:
A Bit of Mathematical History

A Brief Review of What
Mathematics Contained in 1819

If Jefferson had been a professional mathematician and not a statesman, a revolutionary, a philosopher of government and a politician, what mathematics would have been available to him for study? In very broad strokes I will paint the principal features of the mathematical scene as it was in 1819, the date of Jefferson's remark about celestial mechanics. Jefferson was then seventy-five years old.

In contemplating the progress of mathematics, I should point out and it should be kept in mind that mathematics often develops in a helter-skelter fashion and *not* in the "logical" order in which later organizers and writers arrange it. Furthermore, with the passage of time, mathematical notations change, as well as the manner in which a particular mathematical idea is conceptualized and presented.

I should point out further that the presentation that follows in this section is "Eurocentric." This is a dirty word today to a growing number of zealots who argue for a greater historic and didactic recognition of what was developed outside the Babylonian-Greek-European sway. But as regards the manner in which, all over the world, mathematical and technical research is pursued today, the question of what well-established material originated in what land and in what century is fairly irrelevant.

In his famous *Histoire des Mathématiques* of 1758, Jean Montucla defined mathematics in this way:

> C'est la science des rapports de grandeur ou de nombre, que
> peuvent avoir entr'elles toutes les choses qui sont susceptibles
> d'augmentation ou diminution.

[(Mathematics) is the science of the relationships of size and number that can exist between all things that are susceptible of augmentation or diminution.]

Thus geometry as well as all aspects of physics that were quantifiable came within this definition.

This definition, now no longer adequate, would still have sufficed in 1819 when mathematics might very well have been described quite simply as the study of space and quantity. The description of space was the job of geometry while quantity was expounded in arithmetic, algebra, trigonometry and the calculus. These two studies are by no means separate, for geometry, when it talks about concepts such as position, length, area, volume and curvature, involves itself immediately with quantity. Even a concept such as the intersection of two lines, which *prima facie* has no measurement attached to it, can be embedded within a theory of measurement. This occurs in the scheme of analytic or Cartesian geometry made prominent by René Descartes and later writers. On the other hand, disciplines such as trigonometry or calculus, which, if desired, can be pursued in a purely symbolic or formal way divorced from geometry, had origins in geometric questions and draw inspiration from geometric understanding.

The natural integers 1, 2, 3,..., forming the rock bottom basis of quantity, were known in antiquity, and by Archimedes' day (225 BC), their infinitude was proclaimed, clarified, and discussed. The arithmetic of the integers, meaning addition, subtraction, multiplication, and division, was understood. Though not expressed in the algebraic form in which they are currently taught, square roots and cube roots were known in antiquity as well as the solution of quadratic equations. Negative numbers and zero are productions of the Middle Ages, which also saw the beginnings of algebra.

The complex numbers, i.e., numbers of the form $a + b\,i$, where i = the square root of -1, had their origins in the work of the Italian algebraists of the early 1500s; and although the existential nature of such numbers was not clarified until the nineteenth century, many of the formal aspects of their manipulation and their utility to other branches of mathematics were known and promulgated by 1819. For example, in 1799, Gauss demonstrated rigorously that every polynomial of degree n has at least n complex roots, a fact now known as the fundamental theorem of algebra.

Algebra, in Jefferson's day, meant, essentially, the solution of algebraic equations either as individual equations or in simultaneous systems. The letters of algebra, which even today beginners call "unknowns," always stood for numerical quantities of some sort. Shortly thereafter, the palette of "quantities" would be expanded to include a wider variety of mathematical objects, such as quaternions, (William Rowan Hamilton, 1843) that can be combined with each other in some sense and whose combinations exhibit arithmetic-like features and laws.

Determinants are a notion and a symbolism of great utility in the treatment of systems of simultaneous equations, and had their origins in a 1678 notation of Leibniz for dealing with three equations in two unknowns. By 1815, Cauchy was able to announce the fundamental law of the multiplication of two determinants.

Roughly speaking, by Jefferson's day, practically all the raw ingredients of quantity as now understood and practiced, with the exception of higher dimensional or vectorial quantities, were in place, and were accompanied by a rich algebra of manipulative laws and a deep understanding of their intrinsic possibilities.

Geometry was pursued relentlessly by the Greek mathematicians of antiquity (c. 600 BC–200 BC). They knew the straight line, the triangle, the square, and a whole variety of rectilinear figures. They knew the circle, regarded it most highly, and ascribed to it a special role in their descriptions of the cosmos.

They knew the other conic sections as well: the parabola, the ellipse, and the hyperbola. They were familiar with a rich assortment of special curves such as the spiral, the helix, the cissoid, whose properties they studied in depth. They knew the three-dimensional solids with plane faces such as the tetrahedron, the cube, and the three other so-called regular solids. They knew the sphere, the cylinder, the cone, and the other quadratic surfaces. They studied length, area, volume, intersection, parallelism, similarity, tangency. They considered what figures could be constructed by means of ruler and compass alone and placed considerable value (probably aesthetic) on this limitation.

In their approach to geometry, the Greeks introduced the notion of deductive proof. This is a process in which complicated statements are derived logically from certain simpler statements assumed to be self-evident. Deductive proof was an absolutely unique contribution to mathematics. In the *Elements* of Euclid, geometry (and some arithmetic) is presented as a deductive scheme, and Euclid's book, which Jefferson might well have studied, was—and still is—regarded as providing a model for mathematical argumentation and exposition. The notion of mathematical proof or demonstration has played a major role within mathematical methodology, mathematical progress, and mathematical philosophy.

Trigonometry, or the relationship between lengths and angles, is ancient. Spherical trigonometry deals with the measurement of figures on the surface of a sphere and goes back to Menelaeus (c. 100 AD). Claudius Ptolemy (c.150 AD) had well-developed trigonometry and trigonometric tables, though they were not, of course, expressed in the symbolism of 1819.

With the post-Columbian opening up of the world, cartography blossomed and suggested numerous problems tackled by later mathematicians. The abstract mathematical aspect of this craft (as opposed to the compilation of geographical data) concerns itself with how portions of the surface of a sphere can be depicted on a plane surface. It was perceived early that this cannot be done without some sort of compromise and distortion: if some features are preserved others will be lost. Many schemes were proposed (and are still being proposed) and their advantages argued.

The fifteenth and sixteenth centuries, stimulated by the studies and practices of the great Renaissance painters and architects who interested themselves in problems of perspective, saw the beginnings of descriptive and projective geometry. The basic problem there is how to describe three-dimensional objects on a two-dimensional surface, usually a plane. This theory reached high points in Jefferson's period in the work of Brook Taylor in 1715, Monge in 1799, and Poncelet in 1822.

Also within Jefferson's lifetime, there were rumblings of the possibility of alternate geometries, that is to say, geometries that did not obey the famous Fifth Axiom of Euclid. This axiom—considered self-evident (as all axioms are so considered)—asserted that through a point not on a straight line, one and only one parallel line could be drawn. This axiom seemed to critics rather less transparent than the others, and there is a long history of its being subjected to unusual scrutiny.

Gauss in 1813, Lobatchevski in 1825, and Bolyai in 1832 created a so-called non-Euclidean geometry, equally consistent in the logical sense with the classic Euclidean geometry. In so doing, and in the later work of Bernhard Riemann, they both enriched geometry and placed it in an exceedingly ambiguous position. Where originally there had been a single set of geometric truths, the Euclidean, there now emerged an infinite number of geometries, of equal logical status and co-consistent with the Euclidean. With this development, geometry, in its new and enlarged sense, was abstracted away from *a priori* principles that were thought of necessity to duplicate the perceived geometrical nature of the universe. Post-Jeffersonian geometry issued a declaration of independence from physical space. Later, in the days of Einsteinian relativity, the process was turned around: abstracted and generalized geometries were adopted as the mathematical *mis-en-scène* of experiential space.

In the early seventeenth century, two remarkable fusions of space, quantity, and time, each with separate origins, took place. These were later themselves to fuse, and to give rise to a new golden age of mathematics. They were the Cartesian geometry and the differential and integral calculus.

By locating geometrical objects within a numerical grid reminiscent of the labeling of the streets in midtown Manhattan, by referring all geometrical questions to the "addresses" or the "coordinates" of the geometric elements, the mathematicians of the seventeenth century succeeded in reducing geometry to arithmetic and algebra. All questions in geometry could now, in principle, be answered by algebra; and the laborious, clever, but often seemingly accidental, lines of deductive argumentation found in Euclid could be replaced by mere rote computation. This was one of the dreams of Descartes, a dream that was wildly successful, and which, in Jefferson's day, had begun to alter the character of geometry as an independent branch of mathematics.

Responding to the internal pressures of pure mathematics, the differential calculus grew out of the question of finding a general procedure for determining tangents to curves, a problem known in the mathematics of Greek antiquity. The integral calculus grew out of the general question of finding the area enclosed by curves.

Responding to the external pressures of the emerging kinematical and dynamical theories of motion of bodies, developed principally by Galileo and Newton, and formalized and expanded enormously by the French mathematicians of the eighteenth century, the differential calculus grew out of the need to find an appropriate mathematical methodology and symbolism for handling a twofold differencing process to express accelerations. Reciprocally, the integral calculus grew out of the necessity of turning around the differencing process so that the Newtonian equations of motion, initially expressed in terms of accelerations of objects, could lead to information about the positions of objects.

Differentials are analogous to differences while integrals are analogous to sums, and these are the two key ideas of the calculus. The reciprocal relationship between the two, now known as the fundamental theorem of calculus, though implicit in the work of Newton and of Barrow (1630–1677), his professor, was first made explicit by Leibniz in 1675. The basic idea of this theorem is easily enough understood if we restrict our considerations to discrete sequences of numbers and not to continuous functions as the calculus does.

Begin with a sequence, S, of numbers chosen quite arbitrarily, say

(S) 1 1 2 2 3 4 5 4 3 3 1 4 1 4 1 2 6 3.

Now obtain a second sequence (I) formed from (S) by taking its "running sums," i.e., 1, 1+1, 1+1+2, 1+1+2+2, etc. The new sequence obtained in this way is

(I) 1 2 4 6 9 13 18 22 25 28 29 33 34 38 39 41 47 50.

Note that if one subtracts successive terms of (I), one naturally obtains (S) (with the first term omitted): 2 - 1 = 1; 4 - 2 = 2; 6 - 4 = 2, etc.. Using terminology very loosely, the S sequence is the difference, the change, or the rate of change of the I sequence, while reciprocally, the I sequence is the sum or the "integral" of the S sequence. If rectangles with unit length bases are constructed sequentially left to right with the numbers of the S sequence as altitudes, then the accumulated area of the rectangles beginning from the left are given by the values in I.

The sequence of differences, D, i.e., the differences of S, or the differences of the differences of S, DD, are *absolutely vital* to the science of mechanics and appear under the guise of accelerations:

(D) 0 1 0 1 1 1 -1 -1 0 -2 3 -3 3 -3 1 4 -3

(DD)   1 -1 1 0 0 -2 0 1 -2 5 -6 6 -6 4 3 -7

It took several thousand years, abandoning the Aristotelian notions of motion, to arrive at the notion and the importance of acceleration. This was the Galilean revolution and it fed shortly and directly into the Newtonian revolution. The presentation in this paragraph is in its digitalized form.

Throughout the late seventeenth century and the whole of Jefferson's lifetime, the calculus was enlarged and enriched by the labors of many brilliant mathematicians. Perhaps the most brilliant of these was the Swiss mathematician Leonhard Euler (1707–1783)—I like to call him the "Mozart of Mathematics"—whose outpourings in every branch of the subject, pure and applied, were phenomenal both in quality and in quantity. It is quite conceivable that Jefferson had peeked into Lacroix's "*Traité du calcul différential et du calcul intégrale*," which summed up matters as of 1797 and which served as a standard and popular text in the subject for a number of decades.

In Jefferson's day, the calculus, judging from Lacroix's book, included among its major topics: derivatives, series developments, indeterminate forms, extremum problems, tangents to curves, curvature, singular points,surfaces, integration of functions of one and several variables, calculus of variations, integration of ordinary differential equations including approximate methods for doing this, difference equations, differential-difference equations (!) (called today delay-differential equations), the classification of transcendental functions including the elliptic functions.

Roughly speaking, an American student in the year 2000 who takes four semesters of calculus will have covered perhaps one-third or one- half of this material but with many changes of emphasis and presentation.

Basic to the calculus was the notion of function (in the specialized mathematical sense of the term; not in the common meaning of "use" or "purpose"). Though the dynamic quality of the universe was perceived by Heraclitus (sixth century BC) in his famous statement *panta rhei* (everything flows), relations between various components of the universe were conceived by the classic Greek scientists statically, atemporally. Where motion was concerned, as with the motion of the planets, it was always built up from an eternally given, uniform circular motion. As late as the 1660s, the Scottish mathematician James Gregory's notion of a function as that which can be built up from the four arithmetic operations plus root taking plus certain limiting processes, retained a measure of the ancient static rigidities.

The need for accelerations as perceived by Kepler, Galileo, and Newton changed all this. Time enters the picture as the ultimate independent variable to which all must now be referred, and the notion of function as a correlation or a relationship between two sets of variables is born. Lacroix says at the very beginning of his massive three volume compendium of calculus (1797):

> Every quantity that depends upon one or more other quantities is called a *function* of these last; whether one knows or whether one ignores which operations one has to go through to pass from the latter to the former.

A fixed function (or curve), considered in isolation from other functions (or curves), has its own peculiarities or properties. Many of these properties can be formulated in terms of the differential or integral calculus. For example, the problem of where the maximum or the minimum of a function (curve) occurs can be linked to the problem of where the derivative of the function changes its arithmetic sign.

This is easily, but crudely illustrated with the sequences S and D. The S sequence hits peaks of local maxima at the 7th and 17th positions where it has the values 5 and 6 respectively. Now note that the D sequence changes its sign at the 7th and 17th positions (and elsewhere where local minima occur).

In the subject known as the calculus of variations, the "cast of characters" is enlarged from one fixed function to large families of competing functions. Each individual function has a characteristic value associated with it, and the task is now to identify the particular function that optimizes that characteristic value. For example, given two points A and B on the same side of a river, identify a path of minimum length that goes from A, touches the river somewhere, and ends up at B.

The birth of the calculus of variations is usually regarded as having occurred in the year 1696 when the first substantial specific problem was solved. This was the problem of the *brachystochrone*, i.e., a particle is constrained to move along a certain path from a higher point to a lower. Find the path that minimizes the time of descent. The answer is a cycloid, and was found by Newton, Leibniz, and the Bernoulli brothers.

In the year 1744, when Jefferson was one year old, Leonhard Euler published a remarkable systematization of this kind of problem in his book *Methodus inveniendi lineas curvas maximi minimive proprietate gaudentes...*, and with its publication, the theory of the calculus of variations can be said to have been born.

The so-called Euler differential equation, $dL/dy - (d/dx)(dL/dy') = 0$, remains fundamental in this theory.

The development of techniques of computation, currently designated as the subject of "numerical analysis," has paralleled that of mathematics itself and has been fostered by the finest masters of mathematics. This statement may seem strange to the general reader for whom mathematics may appear to be synonymous with computation. The synonymy may have been the case in deep antiquity—even there it is a bit doubtful—and certainly by the time of Euclid, mathematics had developed other goals and interests than those of "mere" computation. If one reads through the pages of Euclid, though quantity is a central theoretical concept, the computational aspect of quantity is hardly dealt with.

The scientists of Jefferson's day had available to them dozens of numerical strategies for such computational processes as the roots of polynomials, the areas bounded by curves, the solution of differential equations. Hot off the press in 1809 and 1810 were Gauss's and Laplace's treatments of the methods of least squares as a method of handling conflicting experimental data.

To ease the labor of brute calculation, the age of Jefferson had at its disposal arithmetic tables, tables of logarithms, trigonometric functions and other specialized functions, some fine drawing instruments, and some very rudimentary digital computational instruments and machines.

Despite these aids, brute arithmetic calculation, though totally mechanical, was time-consuming and tedious in the extreme. It was error-prone, and when the opportunity arose, it was given over as "downstairs work" to duller wits who could do such work while "chewing gum" and dreaming of the silver mines of El Dorado. These features, together with the increasing computational demands posed by the increasing mathematization of life and the increasing complexity of mathematical physics, exerted a pressure to develop fast computation, a pressure that ultimately resulted in the 1940s in the electronic digital computer. The incredible speed and versatility of the computer has revolutionized some aspects of our day-to-day lives and has contributed surprisingly to revised assessments as to the philosophic nature of mathematics itself. There is no doubt that Jefferson, who was mechanically minded and who devised all sort of mechanisms for his home at Monticello, clocks, dumbwaiters, revolving chairs, lap desks, would have been fascinated by the modern computer.

The integers 1, 2, 3,... have always exerted a strange fascination in their individuality. They have often been thought to contain the magic key to various aspects of the universe. For example, Macrobius (c. 450 AD) writes an extensive chapter about the role that the number "seven" plays in the universe. The quincunx is the standard arrangement of the five dots on the "five face" of a die, and the physician Sir Thomas Browne (1605–1682) tells us in his book *The Garden of Cyrus* how the plants in the Garden of Eden were arranged in that manner and what the inner significance of this particular arrangement is.

Even today certain integers carry names that reflect these old opinions. For example, there are even numbers and odd numbers, square numbers, triangular numbers, cubic numbers, prime numbers, perfect numbers, etc. The theory of numbers is the branch of mathematics-and surprisingly one of the most difficult branches-in which integers, combined only by the four operations of arithmetic, are studied in relation to one another. In the third century, Diophantus wrapped up what the ancients knew about this subject, and there the matter rested, more or less, until the study was revived in the seventeenth century. Jefferson's scientific contemporaries would have had two major turn-of-the-eighteenth-century works available to them on this topic, one by Legendre and one, several years later, by Gauss.

As one early (post-classical) achievement of the theory of numbers, we cite the "four squares theorem" conjectured by Fermat around 1640. Every integer, Fermat asserted, is the sum of not more than four square numbers. (E.g., 23 = 9 + 9 + 4 + 1) This proposition remained undemonstrated until 1770, when Lagrange succeeded at the task. Jefferson had known Lagrange personally, and had referred to him as "the greatest mathematician now living." The notorious "Last Theorem of Fermat" (that the equation $x^n + y^n = z^n$, has no solutions for integer $n > 2$, and for positive integers x,y,z) has only recently been demonstrated. This demonstration is beyond the comprehension of the amateur and remains unexamined by most professionals.

Mathematical probability, arising from a desire to provide a rational theory for the way chance operates, whether in gambling or insurance or matters of evidence, grew from modest beginnings in the sixteenth century to a full-fledged subject in Jefferson's day. The first substantial contribution to probability theory was Cardano's (1501–1576) *Liber de Ludo Alea* (*The Book on Games of Chance*). Galileo, Pascal, Fermat, Huygens, James, John and Nicholas Bernoulli, Pierre de Montmort, famous mathematicians all, contributed to the subject. For a while, theories of gambling dominated the scene; later, problems of insurance.

In 1812, Laplace's great memoir on the subject, "*Théorie Analytiques des probabilités,*" was published, advancing both probability theory itself as well as certain techniques of advanced calculus and of early complex variable theory. It is an interesting piece of historical irony that Laplace's "*Mécanique Celeste*" and the "*Theorie Analytique des probabilities,*" the product of one man's thought, contain two philosophical tendencies and methodological programs that are opposed to each other: that of the deterministic event and that of the chance event.

In one of Laplace's papers, (*Sur Les Comètes*, 1813), both points of view are found combined in his attempt to compute the probability that a comet will have an orbit that is parabolic, elliptic, or hyperbolic.

Within two decades of Jefferson's death the deterministic and the random were also combined in the work of the Belgian Adolphe Quételet. A jack-of-all-trades who wrote opera librettos and poetry, a scientist and a scientific "wheel," Quételet (1796–1874) was trained as a mathematician and worked on differential geometry. He learned probability theory by reading Laplace and Fourier and he introduced statistical methods into astronomy.

Quételet is scarcely mentioned in standard books on mathematics, for his work on differential geometry and astronomy is totally overshadowed by that of Carl Friedrich Gauss (1777–1855). Quételet is most noted for his use of statistics in the social areas, the beginnings of which in the modern sense of the term can be traced as far back as the publication of John Graunt's *Observations* (on mortality rates in London), 1660.

Quételet invented the average man (*l'homme moyen*), a notion from which we both profit and suffer. His book *Sur l'homme et le développement de ses facultés, ou essai de physique sociale* (1835), was intended to lay a basis for a "social physics" that would discover and establish quantitative laws of society as solid as the accomplishments of computational astronomy.

Today, the deterministic and the random or chaotic seem each to contain aspects of the other, and this point is very much in the forefront of contemporary scientific exploration and discussion. In Jefferson's day, though, mathematics was often divided into pure mathematics and mixed mathematics (as applied mathematics or mathematical physics was then called). The pursuit of mathematics had not moved away from the pursuit of physics to the extent that it has today; both were seen as part of "natural philosophy." There is no doubt that the most significant link between mathematics and physics lay in Newton's law of universal and mutual gravitation, expressed as a simple functional relationship between masses and distances, supplemented by Newton's laws of motion, and expressed as a system of differential equations involving masses, distances, and time. The working out of the mathematical consequences of these laws, the comparison between what the calculations and what the telescopes revealed, was the subject of Jefferson's comment to Adams.

And Today?
Theorems beyond the Dreams of
Euclid or Euler

How much mathematics is now available? How much new mathematics is being created? One way of answering is to say that mathematics is created at the rate of about 100,000 new theorems per year. Another way is to say that there are now perhaps several hundred sub-specialties of mathematics, partially interrelated, but sufficiently independent that few individuals are deeply familiar with more than one or two of them. Though asserted over and over again by the mathematical establishment, the vaunted unity of mathematics is threatened by its very bulk.

Today's information theorists utilizing hypertext possibilities speak crowingly of unifying mathematics by subdividing it into tiny sub-sub-sub-fields and glueing these subfields together with appropriate computerized links. If one deals merely with raw bytes, the answer is "certainly." But knowledge, experience, wisdom, creative innovation do not proceed from raw—even taxonomized and linked—information alone. The complexity of human experience the possibilities of new experiences, rational and trans-rational, that feed into mathematics are commensurate with the complexity of the whole living population on the face of this earth.

Here is yet a third way of comprehending the immensity of what has been developed. As a frequent reviewer of mathematical books, I am routinely sent review copies of newly published works in the field. On my present schedule, I can handle about one book per month. The bulk of the submissions pile up in my office. I had some floor-to-ceiling bookshelves installed. They are now filled. In desperation, I stacked the books on the floor in front of the shelves. A recent jolt collapsed the pile, and when I looked at it, it reminded me of the pile of logs tossed in through the cellar window when I was about twelve, which I used to split for our furnace. I then had a brief and mathematically sacrilegious picture of myself shoveling the pile of books into a furnace just as the Calif Omar is said to have tossed all the manuscripts in the Library at Alexandria to warm the waters of the public baths. And note this: since I receive books only from a few publishers, the logjam of books in the corner of my office is only a tiny fraction of the advanced monographs that are around.

With the results of 250 years of mathematical research (since Montucla) begging to be wrapped up in a one sentence definition (impossible task!), I like to say that mathematics is the science and the art of quantity, space, and pattern, and of the formalisms and deductive and computational and experiential structures through which its ideas are expressed, applied, and validated. And as though this definition were not long enough, I have found, that as the years go on and as additional possibilities emerge and are publicized, I have wanted to tack on more and more clauses.

I contemplate my incomprehension of the silent and potential infinitude of mathematics, and paralleling Blaise Pascal's emotions, I feel myself filled with both wonder and fright. *"Le silence éternel de ces espaces infinis m'affraie."*

A Fast Forward on the
Adams-Jefferson Question

From December 1998 to January 1999, the John Hay Library at Brown University put on an exhibition of fundamental books on the topic of celestial mechanics. Ranging from Aristotle's (384–322 BC) *De Caelo* and proceding through the works of such scientists as Tycho Brahe, Kepler, Galileo, Newton, Poincaré, Kolmogoroff, to Vladimir I. Arnold's 1963 paper "Small denominators and problems of stability of motion in classical and celestial mechanics," the display contained a vivid record of some of the greatest achievements in the field.

The items were selected by David Pingree and Henry Pohlmann. The rarest of the items from the point of view of the bibliophile, displayed under cloth so as to preserve the beautiful hand coloring, is the *Astronomicum Caesareum* (Ingolstadt,1540) of Peter Apian (1495–1552), which contains "circular slide rules" for determining the longitudes of the planets according to the Ptolemaic system.

In point of fact, the subject of celestial mechanics predates Aristotle by several millennia; it postdates Arnold, and will probably remain on the mathematical research horizon for an indefinite future.

Adams and Jefferson correspond about the stability of the solar system, a problem of celestial mechanics. Recall Jefferson's comment to Adams:

> One of these [papers of Bowditch] impairs the confidence I had reposed in LaPlace's demonstration, that the eccentricities of the planets of our system could oscillate only within narrow limits, and therefore could authorize no inference that the system must, by its own laws, come to an end.

How does this matter stand today, mathematically speaking? The problem, briefly, is this. Given that the planets, their satellites, the asteroids, the comets, the meteorites, of our solar system are attracted to the sun and to each other according to the inverse square law of gravitation, and that their various motions proceed according to the Newtonian system of differential equations, what can one say about the long-run behavior of the total configuration? Will these various bodies stay out of each other's way or will they collide? Will some of them be pulled into the sun? Will some of them eventually escape from the system and fly off into the wide blue yonder?

To answer these and related questions, simplifying assumptions are made: one ignores everything outside the solar system; one reduces the various bodies to point masses; one forgets about Einsteinian relativistic effects. And even with these and with many other simplifications, the mathematical problem is formidable. (The problem is already formidable with only three bodies, say the Sun, the Earth, and the Moon.) Hundreds, perhaps thousands of mathematical papers by the most brilliant celestial dynamicists have been written. Since the advent of the electronic computer, numerous computations have been executed in which time is carried forward millions of years.

What say these computations? I limit my answer to certain fairly recent results described in an excellent survey article by Jacques Laskar entitled "The stability of the solar system from Laplace to the present" that can be found in Volume 2 of *The General History of Astronomy*, Taton and Wilson, eds., Cambridge University Press, 1995. Laskar's article is an appendix to a more extensive history, "The Golden Age of Celestial Mechanics," by Bruno Morando.

In 1988, Sussman and Wisdom, restricting themselves to the outer planets and computing over a period of 875 million years, found Pluto's motion to be chaotic. This means extreme sensitivity to initial conditions, and so prediction via computer beyond 400 million years is impossible. Shortly thereafter, Laskar himself worked with the inner planets, Mercury, Venus, Earth, Mars, and employed a different numerical formulation than Sussman and Wisdom, one that contained 150,000 terms and yielded only the average motions of the planets.

The result was surprising. For the large planets, the result was a regular motion, but for the interior planets the behaviors of the trajectories was chaotic. Initial uncertainties were found to increase by a factor of 3 every 5 million years, preventing all prediction beyond 100 million years.

Celestial mechanics, which Laplace erected as the model *par excellence* of predictable science, has shown its limits. A new formulation of the problem of stability is imposed on us today. The solar system, we have shown, is unstable; it is a matter of now knowing with precision the effects of the instabilities for a time comparable to the system's age. To do this it is necessary to study globally all the neighboring trajectories, and thus follow the way opened a hundred years ago by Henri Poincaré. A better knowledge of the ensemble of these motions will not permit us to predict whether a catastrophic event, such as a sudden increase in the Earth's orbital eccentricity, will actually happen in the next billion years; it will authorize us only to say whether such an event, within such a period of time, is possible or not.

Millions of years? Billions of years? Human time is not measured in such units of time. And perhaps that is why humans seem to be in such a rush.

For a much more technical and theoretical update, the reader is referred to Juergen Moser's article "Dynamical Systems—Past and Present," pages 381–402 of *Procedings of the International Congress of Mathematicians, Berlin, 1998, Vol. I: Plenary Lectures and Ceremonies.*

In Moser's article, the problem is formulated in this way:

> The stability problem for Hamiltonian systems [William Rowan Hamilton (1805–1865), Irish applied mathematician] is an old unsolved problem which fascinated many mathematicians in the past... This is modeled by the N-body problem where N masspoints of positive masses m_j move in Euclidean space R^2 or R^3. One asks for bounded orbits avoiding collisions.
>
> The question of stability requires not only finding single orbits [of this type], but an open set of such solutions [i.e., a family of nearby orbits], and accounting for the imprecise knowledge of the initial values...
>
> In spite of modern advances in this field, this is still an *open problem*!

We have thus seen that in his plough, his cryptography, and in his response to Adams about celestial mechanics, Thomas Jefferson put his finger on what have turned out to be very difficult, abiding, and significant problems of applied mathematics. Not bad for an amateur!

Part X

Mathematics and the French Enlightenment

Voltaire and Newton

In the book that Bernhelm and I sketched out orally and in many subsequent e-letters, we fantasized that the Ghost of Thomas Jefferson had asked us to bring him up to date with respect to mathematics; to tell him what mathematics has accomplished in the two and a half centuries since he studied Hutton's popular *Course of Mathematics* or browsed in Laplace; to tell him how and in what ways it has molded the physical and the mental world of the twentieth century; what are its prospects for the twenty-first; whether it has contributed to his ardent hope that the forward movement of science is linked to the notion of freedom, and whether it has been instrumental in throwing off despotisms.

A full description of the intellectual ideas of the eighteenth century, on which Thomas Jefferson was nourished, grew and to which he contributed, would require a dozen thick volumes. By way of "Jefferson prep" as the nurses in a hospital might say, I will introduce several historic characters and tell a few stories that relate them with mathematics. I'll start with Voltaire (Francois-Marie Arouet, 1694–1778).

Voltaire's *Candide*, a great classic and a book that raises the eternal question of why evil exists, was in my house when I was a kid. We had a nice edition in a lovely binding that sported somewhat risqué woodcuts done in the Art Deco manner. I read Candide over and over again. At first I was much too young to understand what was going on or what its point was. Candide was a simple, decent, credulous fellow; Dr. Pangloss, a philosopher who insisted that this was the best of all possible worlds despite evidence to the contrary; the beautiful Cunégonde, the object of Candide's most ardent desires. A few years later I finally understood what was going on and despite my own strong tendencies towards satiric formulations, I found the dozens of fierce but fictional atrocities in *Candide* approached so closely those I had been reading about in the newspapers that the satire was disturbing.

Now, if the reader looks up Voltaire in a dictionary, it will say that he was a poet, a dramatist, an essayist, and a historian. That much I've known for years. But Voltaire was also a scientist and a man who had a prolonged *amitié* with a mathematician, Emilie du Châtelet, a fact not reported by most dictionaries and given short shrift by the longest of encyclopedia articles.

Voltaire was an advocate of the Newtonian system as opposed to the Descartian vortices (which the French had long clung to), and collaborating with Emilie, he wrote a long description of Newton's ideas (*Eléments de la Philosophie de Neuton* (sic), Amsterdam, 1738), rarely mentioned in books on the history of mathematics, but which, one of my physicist friends tells me, is the best-written popularization of Newton he's ever read. There are lots of geometrical figures in this book, but not a single equation. I doubt if Voltaire knew much mathematics. (Emilie helped him over the rough spots.) In one place in his book he makes a mistake in his definition of the sine of an angle, and speaks of the concept as though it were far beyond the knowledge of the educated laity. Nonetheless, he once wrote that the calculus was concerned with the accurate computation and measurement of quantities that don't exit. (I wonder whether he'd read Bishop Berkeley's *The Analyst* (1734) in which the differentials of calculus were called "ghosts of departed quantities.")

Emilie du Chatelet:
Divine Mistress

The sobriquet "The Divine Mistress" was given to Gabrielle Emilie le Tonnelier de Breteuil by Voltaire, who was her lover. The "du Châtelet" was added on after she married the Marquis Florent-Claude de Châtelet. She was handsome, though as a child she had been an ugly duckling. She was tall. She was an aristocrat with not too much money for her aristocratic and reckless tastes. She was a clothes horse. She wore deep décolletage. She loved expensive, dazzling jewelry, carriages, palaces. Her lovers, particularly Voltaire, who was an enormously successful writer made wealthier by royal allowances and prudent investments, enabled her to make a great splash.

She was proud, vain, passionate in a dozen different ways. She loved partying and dancing. She loved to act in plays, to sing. She was addicted to gambling. She often placed bets at breathtakingly high levels—only royalty would have ventured such sums—and she won and she lost. Occasionally Voltaire would have to bail her out of her deep gambling debts.

She was born in 1706 and married an army general who became grand marshall and who was frequently away with his troops on the borders of Germany. She had a series of lovers, of whom Voltaire was the most significant. She lived with him for a decade and a half on her husband's property at Cirey-sur-Blaise. Cirey was partly a laboratory. The whole of aristocratic or intellectual Europe was abuzz with gossip about the couple at Cirey. Their house guests wrote detailed descriptions of the goings on under their very noses.

Emilie had brilliant social skills, but was a moral and an emotional idiot. Those were the times, one might say. On the other hand, one might also say that apart from her absolutely total and almost neurotic dedication to Voltaire, she was a completely independent woman. She wrote to Frederick the Great of Prussia when he deliberately did not invite her to Sans Souci along with Voltaire,

"I am in my own right a whole person, responsible to myself alone for all that I am, all that I do."

And there was another side to her. Listen: how would you react if you heard that Madonna (of popular culture) was writing informed and serious stuff about Stephen Hawking's theory of cosmology? Emilie studied and discussed mathematics with Alexis Claude Clairault and Pierre-Louis Moreau de Maupertuis, both names in the books on the history of mathematics. She hung around the Café Gradot, where scientists, philosophers, and academicians foregathered and where women were not supposed to be found. She was a whiz at languages.

She talked constantly, often on high-flown physical or metaphysical topics. The women hated her for this; her husband found her talk soporific. She read voraciously. She wrote steadily, in low-cut gowns and with ink-stained fingers. She had the knack of explaining difficult concepts clearly. Her house at Cirey was filled with laboratory experiments she was constantly performing. It was scribble, scribble, scribble, all day long, for both her and Voltaire.

She translated Newton's great work, the *Principia Mathematica*, into French (still the only translation and reprinted recently), introducing French scientists to Newton's gravitational theories. (Actually it was only published posthumously, prepared for publication by Clairaut.) She collaborated with Voltaire in writing his *"Eléments de la Philosophie de Neuton."*

She translated Vergil's *Aeneid* into French, a translation that was used in French schools for a century. She wrote an exposition of Leibniz's philosophy and science in three volumes. If it hadn't been for Emilie's sympathetic treatment of Leibniz, the world would probably never have enjoyed Voltaire's *Candide*, a spoof on Leibniz's optimistic, "This is the Best of All Worlds" philosophy. Nor, for that matter, Lenny Bernstein's musical *Candide*.

There is a portrait of Emilie done by an unknown artist at an unknown date, showing her beautifully gowned, a rouche and bow around her neck, looking up from a pile of scientific manuscripts, compass in hand and an armillary sphere in the background.

In 1744, she wrote *Traité sur le Bonheur*, a book on how to live right. It became a best seller and added to her already substantial notoriety. Only a portion of her precepts would now pass muster with today's pop, Sunday-supplement psychologists.

She considered herself a genius. So did Voltaire, who wrote bombastically in a dedicatory poem to his *Philosophie de Neuton*:

"Tu m'appelles à toi, vast et puissant génie
Minerve de la France, immortelle Emilie
Disciple de Neuton, et de la vérité
Tu pénètres mes sens, des feux de ta clarté"

[You call me to you, prodigious and powerful genius
Minerva of France, immortal Emilie
Disciple of Newton and of truth
You penetrate my senses with the fires of your clarity]

Some considered Emilie a poseur. Some considered her a witch. (This was in a day before rationalism had set in seriously, and to call someone a witch was an option available to anyone who was jealous.) In any case, whether or not Emilie du Châtelet makes it into the standard histories of mathematics, she was a formidable European phenomenon.

She had three children and, at the age of forty-three, made pregnant by her last lover, the Marquis de Saint-Lambert, she died in childbed.

Voltaire's character can be read in his remark that the child should be classified among her "Miscellaneous Works."

Diderot, d'Alembert, and Rousseau

A sharper way, perhaps, of understanding the background of some of Jefferson's opinions is to describe the beliefs of three men—particularly as they relate to mathematics: d'Alembert (1717–1783), Diderot (1713–1784), and Rousseau (1712–1778). These men were among the leading thinkers of the French Enlightenment. I begin with Diderot.

Diderot, the energizing spirit of the *Encyclopedia* project, was an atheist, a materialist, and a determinist. He was a polymath and a philosopher whose philosophy was tremendously individualistic. He was responsible for the often quoted remark that he would like to see the last king strangled by the guts of the last priest; and yet he was also the man who, at Catherine the Great's request, had the guts to go to St. Petersburg and tell her to her face that Russia needed liberalization and to write a tract instructing her exactly how to do it. It came to nothing, of course. The execution of Louis XVI gave her the shakes and she clamped down.

Diderot's *Encyclopedia* (*Encyclopédie ou Dictionnaire Raisonné des Sciences, Arts, et Métiers*) was conceived to fill two explicit functions: to serve as a dictionary, and to exhibit the schematic linkage of all knowledge. The prospectus for the *Encyclopedia* pointed out that the very word "encyclopedia" contains two roots: one for the circling, and the other for institution or science. The *Encyclopedia* served another function that was not acknowledged publicly: to present a philosophical position that was at odds with the established view. For this reason, in the twenty or so years in which it was in production, its writers were harassed and threatened with imprisonment, individual articles were often self-censored or censored by the printer, and individual volumes were proscribed, banned, and burned by the public hangman. In later years,

Diderot's other work was censored by relatives and bowdlerized even by Goethe (who was, in fact, one of Diderot's great admirers).

The *Encyclopedia* organized the totality of human knowledge and understanding into three categories that were labelled "memory," "reason," and "imagination." Under "memory," one finds history, sacred, ecclesiastical, civil, and natural. Under natural history we have the uses of nature: various arts, trades and technologies. Under "reason," one finds philosophy, which embraces metaphysics, etc., theology (as the science of God), logic, ethics, mathematics, and what we would now call the physical and biological sciences. "Imagination" is identified with poetry. It splits into the sacred and the profane, then splits further into narrative, dramatic and allegoric.

Let us now take an in-depth look at the organization of mathematics. First of all, it comes under the rubric of "the metaphysics of bodies, or the general physics of extension, impenetrability, movement, the Void, etc." Mathematics divides into two parts; I: Pure mathematics; II: Mixed (i.e., applied) mathematics. Here is the scheme.

I: Pure Mathematics

1. Arithmetic
   Numerical
   Algebraic
      Elementary
      Infinitesimal
         Differential
         Integral

2. Geometry
   Ordinary
   Transcendental
      Finite
      Infinitesimal

3. Algebra
   Finite
   Infinitesimal
      Differential Calculus
      Integral Calculus
      Exponential Calculus

II: Mixed Mathematics

1. Mechanics, Statics and Dynamics
2. Astronomy
3 Optics (including perspective)
4. Acoustics
5. Pneumatology [i.e., gases]
6. The Art of Conjecture, and Hazards [Probability]

One of the things that strikes the contemporary mathematician is that logic is placed outside of mathematics. Today, logic, at least mathematical logic, is considered a branch of the subject, often a fundamental branch, but it suffers from the paradoxical situation that most practising mathematicians need to know very little of its technical content.

Of the three men mentioned, Rousseau is certainly the most famous. Rousseau is the "back to nature" man, the man who sang the praises of "the noble savage" and proposed the notion of the "social contract." He wrote that human ideals derive from love and that, while mankind is good by nature, it has been corrupted by civilization. Perhaps more than most of us, Rousseau was a bundle of contradictions, and more than a bit of an unpleasant oddball.

In the words of historian Peter Gay, he

> was a playwright who inveighed against the theatre, a moralist who abandoned his children, a religious philosopher who changed his confession twice for dubious reasons, a libertarian who could not get compulsion out of his mind, a deist who accused his fellow deists of irreligion, a professional celebrant of friendship who broke with everyone.

D'Alembert, the illegitimate son of two aristocrats, was a professional mathematician and certainly one of the most distinguished of his day. His work on algebra, partial differential equations, and mechanics has given his name to numerous topics in these fields. On the side, d'Alembert was a philosopher of sorts, a religious skeptic, an anti-Jesuit, a defender of tolerance, and a man with strong views on the role of mathematics in science. He was responsible for most of the mathematical and scientific entries in the *Encyclopedia*.

Mentioning two of these men, Jefferson wrote:

Diderot, d'Alembert, d'Holbach, Condorcet, are known to have been among the most virtuous of men. Their virtue, then, must have had some other foundation than the love of God. – Letter to Thomas Law, June 13, 1814.

Rousseau knew rather little mathematics. Diderot, on the other hand, knew a fair amount for a layman. He even wrote a couple of books on elementary mathematical topics, and in his forty-ninth year and on and off for the rest of his life, he was hooked on the problem of squaring the circle. (In his day, the impossibility of the task had not yet been settled, and he thought he had found a way to do it.)

Diderot knew mathematics, but only up to a point, and in a "helpless longing for the old comprehensible scientific style," he denounced the austere mathematical language of Newton's *Principia Mathematica* as the "affectation of the great masters," as "the veil" that scientists "are pleased to draw between people and nature." And he wistfully reiterated the old ideal of the universal cultivated man:

> Happy the geometer in whom a consummate study of the abstract sciences has not weakened the taste for the fine arts; to whom Horace and Tacitus will be as familiar as Newton, who could discover the properties of a curve and sense the beauties of a poem. - Peter Gay, *The Enlightenment*, vol. 2, p. 158).

While from time to time these three men were friends and colleagues, they had their fallings out, and their views on the nature and position of science were by no means identical.

What did Diderot think of mathematics? In *On the Interpretation of Nature* (*Pensées sur l'interprétation de la nature*), Diderot questions the grand pretensions of mathematics. He is rather more intrigued by the natural and experimental sciences, chemistry, electricity, technology, that were then opening up in a marvelous way. Perhaps this is his perception in view of the fact that his father was a manufacturer of forks and knives. The future lies with science and technology. Methodologically, he stresses "conjecture," by which he seems to mean an intuitive and experimental grasp of the way that nature works.

Diderot says that mathematicians have accused other scientists of being metaphysical; now, other scientists are claiming as much of mathematicians.

Following Buffon (*Histoire Naturelle*), he asserts that the truths of mathematics are simply the truths of definition. As regards (what we have now come to call) mathematical modeling, he reasserts Buffon's statement that it makes suppositions that are contrary to nature, stripping "an object of most of its qualities and making of it an abstract entity that has no resemblance to reality."

Mathematics, said Diderot, is grand but is played out and exhausted as far as utility is concerned. If mathematics were so all-revealing, he says, "what purpose would all those deep theories of celestial bodies and all those enormous calculations of rational astronomy serve, if astronomy cannot dispense with (James) Bradley and (Pierre Charles) Le Monnier observing the heavens?" Mathematics is about to be dethroned:

> This science will come to an end suddenly where the Bernoullis, Euler, Maupertuis, Clairaut, Fontaine, and d'Alembert have left it. It has set up Pillars of Hercules beyond which there is no going. In the centuries to come, their work will persist even as the Egyptian pyramids with their masses full of hieroglyphs awaken in us the awesome idea of the power and resources of the men who raised them.

D'Alembert's attitude is rather different: he presents a vision of the unity of knowledge, all linked in a chain of propositions. As yet, we have only a few links in this great chain. He asserts that there are only two sources of certain knowledge: (1) The knowledge of our own existence; (2) The truths of mathematics. This assertion by itself would have put him at odds with the ecclesiastical authorities: where does divine revelation fit into this scheme?

In his "Preliminary Discourse" to Diderot's *Encyclopedia*, d'Alembert reduces the universe to its component parts and then builds it up again from those parts. He starts from the perception of the concrete matter in the universe: its extent, its impenetrability, goes to mathematical abstractions and then back to the concrete via mathematical theories of matter. It is revealing to quote his words at some length.

> Geometry deals with the properties of extension as regards shape.
> Since shaped extension presents us with a large number of possible combinations, it is necessary to invent some means of achieving those combinations more easily; and since they consist chiefly in calculating and relating the different parts of which we

conceive the geometric bodies to be formed, this investigation soon brings us to Arithmetic or the science of numbers.

Arithmetic leads us to algebra which is "the science and art of denoting numerical relationships." Then, through

> the continual generalization of our ideas, we arrive at that principal part of mathematics and of all the natural sciences, called the Science of Magnitudes, in general. It is the foundation of all possible discoveries concerning quantity, that is to say, everything that is susceptible to augmentation or diminution.
>
> This science is the farthest outpost to which the contemplation of the properties of matter can lead us, and we would not be able to go further without leaving the material universe altogether. But such is the progress of the mind in its investigations that after having generalized its perceptions to the point where it can no longer break them up further into their constituent elements, it retraces its steps, reconstitutes anew its perceptions themselves, and little by little, and by degrees, produces from them concrete beings that are the immediate and direct objects of our perceptions. These beings which are immediately relative to our needs, are also those which it is most important to study. Mathematical abstractions help us in gaining this knowledge, but they are useful only insofar as we do not limit ourselves to them.
>
> –From the translation by R. N. Schwab.

I have sketched the views of three Enlightenment figures, Rousseau, Diderot, and d'Alembert, mainly as they dealt with science and mathematics. Three men and three opinions. What, then, was common them? Here is the conclusion of Henry F. May, in the *The Divided Heart* (p. 164):

> The Enlightenment was not necessarily, for instance, optimistic. It was not always rationalistic. It was not always empirical or pragmatic in method. About the only common content that I have been able to find can be put into two very simple statements: First, by definition, the Enlightenment consisted of those who believed that the present age was in some sense more enlightened than the past, that people had become better able to

understand the universe. Second, Enlightened people believed that this understanding was best achieved through the use of the natural faculties of the human mind, and not by reliance on either revelation or mystical illumination.

Updating the Ghost of Mr. Jefferson

The author addresses the Ghost of Thomas Jefferson:

Respected Shade, dear Ghost of Hopefully non-Departed Qualities: how on earth can I carry out the plan that Bernhelm and I sketched? It is much too vast.

An update of mathematics? In a sense, I am the update: Bernhelm and I and thousands of mathematicians over the intervening centuries and around the globe who have worked to teach, to develop and to apply the subject. Our jobs would not have existed in your day. Since you were a student at William and Mary, millions of pages of research mathematics have been written. How can one select out of this mass of material the heart of the matter? Or even agree that there is a unique heart? Should I tell you about very specific advances or should I try to be very general?

I would like you to know the very definition and the scope of mathematics has widened tremendously since you were a student. In those days, mathematics was seen as the study of space and quantity. Today, mathematics might be defined as the study of the interrelationships between quantity, space, and pattern, as perceived intuitively, and the abstract symbolic, deductive, and computational structures that have been devised to deal with them. The emphases on pattern and structure have been major conceptual changes. The increased emphasis on deduction has been the major methodological change as regards pure mathematics. I advance this definition in the full knowledge that the contents of a million or so of advanced books cannot be summed up in a few words.

New mathematics and applications of both old and new mathematics are constantly being created. All physical sciences, some social sciences and some graphic arts have tended towards expanded mathematizations. The applications of mathematics embrace description, prescription and prediction, and these are aspects of a general methodology that has come to be called mathematical modeling or simulation. Prediction can lead to actions to be taken in response to the natural world that is "out there." When mathematical prescriptions (e.g., I.Qs; cost-of-living indices) are acted upon—a process that some have called formatting—they can create new social and economic worlds of their own.

Since you loved gadgets of all sorts and devised a few of your own, you will be pleased to hear that the computer has become an indispensable tool for the transformation of mathematical descriptions to objects and to actions of practical utility. In fact, the computer is no mere gadget. It is a revolutionary instrument that has enabled the formal and hopefully, rational manipulation of mere symbols to create mechanisms of control, organization, information, perception, inference, reaction; in short, to replicate and in some respects to go beyond many of the functions of the human brain.

"How goes the United States of America; how goes the Enlightenment?" you asked me when we first made contact. And I answer that all of us here in this country and in the "advanced nations" are children of the Enlightenment and without Enlightenment values we would find life unbearable. But I must add that the legacy of the Enlightenment (I quote from the Foreword to *Construction and Constraint* by Vaughn McKim)

> ... has proven to be a deeply troubling one. The outstanding thinkers of this movement were unremitting in their condemnation of the many forms of prejudice, myth, and superstition which they held responsible for the unsatisfactory social and material conditions of their time. Yet these same individuals evinced a remarkable faith in human rationality and in its handmaids, the "new sciences," believing that once unfettered, human reason would not only conquer the physical world, but usher in an unprecedented era of justice, equality, freedom and happiness.
>
> Intimate acquaintance with the social and political history of the nineteenth and twentieth centuries is scarcely required to realize that the Enlightenment's optimistic prognosis for Western society has not been borne out.
>
> The Enlightenment failed to envision what the real consequences of seeking to rationalize narrowly every dimension of human experience must be. But despite skepticism about the

benign role of rationality in politics and the shaping of society, one plank in the Enlightenment program seemed secure, namely, its conviction that science embodies the capacity to transform our understanding and control of nature, and that it possesses this capacity by virtue of exemplifying human rationality in its purest form.

The purest elements of scientific rationality are often thought to be exhibited by its mathematical formulations. Thus mathematics, and with it, all the inner deductive apparatus of mathematics, are *the prime exemplars* of rationality, and the "reality" which is created by mathematics is often thought to be totally objective: "free of any taint of personal bias, ideology, or gratuitous metaphysical prejudice."

Dear Ghost, you were well up on the classical and contemporary philosophers. Traditional philosophies regarded mathematics as the one area of human thought and experience where absolute certainty reigns. John Henry, Cardinal Newman, a young man in the 1840s, was bold enough to proclaim that the truths of mathematics were more certain than the claims of dogmatic theology. You, Respected Ghost, may raise your eyebrows high on hearing that numerous developments of the past century have shown that this certainty has only a local character. Within the subject itself and in its interaction with the outside, mathematics is laced with ambiguities of many sorts.

The spirit of Positivism, born of the Enlightenment, asserted the logical truth of mathematics but also divorced mathematics from ethics and from theology. You, by contrast, espoused an older, pre-positivistic view, if we can attribute moral content to your notion of freedom, when you expressed the hope that the progress of science and mathematics would advance the cause of freedom.

Nonetheless, you may be shocked to learn that mathematics displays both strengths and weaknesses and that the pursuit of mathematics can be separated from human feelings and aspirations only at a great price. The prescriptive (formatting) power of mathematics in our everyday life is great, but there is a cost. In the words of Hubert L. Dreyfus, total reliance on "computational rationality" weakens the intuition, and in the course of this, "wisdom becomes an endangered species of knowledge."

The age of technocratic Enlightenment, Mr. Jefferson, is still here. We may be "post-industrial," but we are not yet post-rational. And the computer, that mathematical instrument par excellence, with its simulation of mechanized rationality, through its acolytes who promote such rationalities, promises that the changes civilization has experienced from the Enlightenment to the present will be as nothing compared with the changes yet to come. This may very well be the case. Yet those of us who are inclined to worry, do worry.

Nearly a half-century ago, philosophers Horkheimer and Adorno wrote that if the Enlightenment does not accommodate reflection on the destructive aspect of the progress it engenders, then it seals its own fate and ours. If this was true in a day when only low- to medium-level technology was available, it is far truer in the present generation of high and super high technology. And, paradoxically, even as the center of gravity of philosophic opinion has changed drastically over the last half century, technological change, which was legitimized and which was blessed with a metaphysics and by a philosophy now abandoned, has rushed wildly forward, even as the waters behind a collapsed dam, and has almost outrun the possibility of any decent philosophic system catching up with it.

While it seems certain that the program of the mathematization of the world will continue for the next generation with unspent energy, the reasonableness of this program cannot be demonstrated in advance. Its pursuit is an act of faith that characterizes our particular century and our civilization.

Perhaps I should now tell you some new applications of mathematics in areas in which you had particular interest: law, government, diplomacy, war, and politics. But first, a quick and easily understood view of the extent to which our civilization is now mathematized.

Part XI

Applied Mathematics of Jeffersonian Interest

One thing that has always impressed me in my reading about the thinkers of the eighteenth century is that they were interested in EVERYTHING.

What interested Jefferson? Everything: from architecture to zoology. Such is hardly the case today; with the specialization and over-specialization that currently exists, if a scholar, Smith, should perchance venture across two specialized fields, one writes of Smith (on the backs of dust jackets and in book reviews): "Smith is a true Renaissance Man."

I have thought about some of the the applications of mathematics that might have interested Jefferson, that are of "presidential concern" and in the present chapter I will describe two of them: law and war. But first, a panoramic view.

What's in a Can of Peaches? or
The Mathematization of Our Civilization

Dear Ghost, we are now living in a civilization that is far more mathematized than when you held sway in the White House. It was said that if you wanted to see the monument to the memory of Christopher Wren (1632–1723), the mathematician and architect of St. Paul's Cathedral in London, just look around. (*Si monumentum quaeris, circumspice!*) As a mathematician, Wren was noted for his work on the rectification of the cycloid and on the one-sheeted hyperboloid. Mathematics was certainly implicit in architecture from the earliest days, and the theories of such architects as Palladio made a great impression on you when you had your architectural hat on.

If you were to ask me, more generally, where in modern everyday life, can mathematics be found, the answer, again, would be the same: just look around. You will find it everywhere. Or will you? The trouble is that a good deal of the mathematics will be "hidden from view" in the sense that you probably will not realize that it was there. You must dig beneath the surface.

Just prior to writing this section, I opened up a can of peaches for lunch. The amount of mathematics that was suggested to me by a simple can of peaches was extraordinary.

Try this experiment. Take a can out of your cupboard and set it on the kitchen table. (I'm not sure whether canned goods were around in your day. Was food preservation exclusively by drying, curing and pickling?) Well, whatever. Step back a few feet, look at the can and ask yourself the question: what mathematical ideas does it suggest to you, either directly or by extension? Come closer, pick up the can, look at it closely and ask the same question. Finally, open the can up, pour the contents into a bowl, and ask the question still a third time. What follows now are my free-associative thoughts as I performed this experiment.

From afar, what struck me principally was the geometry of the can. It is cylindrical. Most, but not all cans of food are cylindrical. The can of peaches has linear dimensions: its height and the radius of its cross section. These quantities remind me that the can also has surface area and volume. More than this, a can has moments of inertia and products of inertia, linked together in a 3 by 3 matrix or tensor that would play a role if the can had accidentally tumbled off the shelf and if I wanted to analyze and then compute its trajectory. Perhaps the dynamics of a falling can of peaches is not terribly significant, but the dynamics of a satellite, tumbling in space, and whose tumble is governed by the same Newtonian laws of motion would surely be more significant. (Technically speaking, a coupled system of six second-order differential equations.)

Coming back to earth, I recalled a problem that every student works out in the first weeks of elementary calculus: for a given volume, what dimensions should a cylindrical can have in order to minimize its total surface area?

The answer is that its height ought to equal its diameter. You don't need calculus to arrive at this conclusion; just a bit of experimentation with a hand-held computer—better still, a hand-held computer that has a graphics display of functions—will lead you to the answer. I wondered, further, whether there was any real utility is this knowledge, given the fact that most cylindrical cans on the supermarket shelves, including my can of peaches, were not of these relative dimensions. What, then, is the relationship between the knowledge that teachers inculcate, the problems that students drill on, and the larger world of real experience?

I wondered, generalizing, what the answer would be to the problem of finding the minimum surface area for a given volume if the cross section of the can were an ellipse of a fixed size. I wondered how the cylindrical cans would fit together on the shelf most economically, and I knew that geometrical problems such as this, called packing problems, were often unusually difficult to solve for assemblages of three-dimensional objects.

But geometry is not all measurement. There is also pattern. The cylinder exhibits symmetries: one can rotate it about an axis that joins the centers of the top and bottom circles, and the cylinder comes back on itself or occupies the same space. This would not be true if the cross section were anything other than a circle. One can reflect it across the plane that bisects this axis perpendicularly and it comes back on itself. This would not be true of the plane that bisects perpendicularly the major axis of an egg. An egg has symmetries, even though one end is larger than the other, but they are not those of a cylinder.

Now I came even closer. I picked up the can and looked at it carefully. I looked at the label and at the part of the surface that was not covered over. I was amazed: I counted thirty-six separate numbers. Not just digits, but complete numbers. These numbers conveyed different sorts of information: price, weight, date, lot number, the zebra stripe, an 800 telephone number for free product information and complaints, street address, zip code, dietary information, recipe quantities, just to name a few.

I thought about a few of these, one after the other. Many of the numbers are simply for purposes of identification: lot numbers, street address numbers. Some numbers suggested deeper things to me. "For maximum goodness, sell by May 15th." This admonition involves the calendar, and a calendar is a complicated arithmetic-astronomic arrangement, hammered out and refined over a period of centuries. The last calendar adjustment that took place in America was in 1732. Within the current scheme, to figure out what day of the week July 4th, 2736, will fall on, takes a fair bit of number theory, considered elementary by number specialists but not so elementary to the general public. Although the formula is easily chipified, the famous German mathematician Carl Gauss considered it worthwhile to work it out in symbols.

But there is not just one calendar in operation in the world, there are many. There is the Chinese calendar; there is the Muslim calendar, there is the Jewish calendar, there is the calendar of the Eastern Catholic Churches, to name a few. Some are solar, some are lunar, some are luni-solar. Some have leap days, some have leap months. And there have been proposals for "world calendars" that even out some of the irregularities that result from the incommensurable astronomical periods of the day, the month, and the year.

Consider the 800 telephone number to call for product information. Now if there was ever a technology that is mathematized, it is the telephone business. The same can be said of all the various modes of electronic communication. For many years, prior to the breakup of the Bell System by the courts, the Bell Telephone Laboratories in New York City and Murray Hill, N.J., was one of the principal industrial producers and consumers of mathematical research in the US; and hence, one of the major employers of mathematical talent. The mathematics of switching systems is itself a major chapter in applied mathematics.

The label on the can of peaches mentioned the "quality" of the peaches, and something about the size. The quality was listed as AAA, and I wondered how the quality was defined: whether it had a numerical definition or whether it was a subjective matter.

Size reminded me of an experience I had in Chico, California which, at the time, was the US center of kiwi production. A local packing house used a machine to size the fruit automatically. The machine was misbehaving, throwing fruit into the wrong bins, and a mathematician from the local university was called in to fine-tune the sorting algorithm. He had first to find a mathematical description of a kiwi from the commercial point of view.

Enough about the label: I removed the top of the can, and I poured the contents into a bowl. The heavy syrup-peach mixture flowed nicely into the bowl. Give a description of the flow? Well, for that you must consult a specialist in what is called suspension or slurry or multi-phase fluid flow. A description via differential equations is by no means an easy matter.

I tapped the empty can. It emitted a musical note. I recalled that some musical groups played on similar primitive instruments. What kind of a sound? This question is a much studied problem in the theory of the eigenvalues of partial differential operators (i.e., the frequencies and amplitudes of vibration). While the vibrations of a tin can may not be of the greatest concern either to the fruit packer or to the consumer, the vibrations of a bridge, an airplane, a space station, or of the human heart can be of vital concern.

My reverie ended. It had lasted long enough. And I didn't even try to describe the molecular or atomic structure of either the peaches or the can. I didn't even get to the economics of the can, although money remains the first and most pervasive mathematization of civilization.

The can of peaches thus contains a world of mathematics. So does a piano, a road map of Denmark, a gas bill, an automatic camera; a state-of-the-art hearing aid. All are macrocosms in which a mathematical microcosm resides. Every specialist, whether of physics, chemistry, biology, sociology, history, economics, architecture, ecology, could have reveries somewhat different than mine, reveries that would interpret a can of peaches along the lines of their individual specialties. Would it be possible—as with DNA—to reconstruct the main features of our present civilzation from a simple can of peaches?

The world of mathematics may not be the only world contained in the can of peaches, but it is one that is deeply hidden from the view of the public. I often tell teachers that a good way to achieve an enhanced awareness of our mathematized world is to ask a number of questions:

(1) Can you identify hidden mathematizations?

(2) What is the history of these mathematizations?

(3) What would life be (or what was it) without them?

(4) How is a mathematization validated socially, i.e., why does the community install it and keep it in place?

(5) Can you suggest or anticipate new mathematizations?

It is of the greatest importance, Mr. President, as part of a continuing examination of the dialectic of the Enlightenment, that we achieve an understanding of the nature of the mathematical discipline, and why and in what sense mathematicians believe in its integrity.

It is no easy task to arrive at this understanding. Mathematics has entered into more and more phases of our daily lives in a deep and seemingly permanent way. As further evidence of this penetration, consider a few more examples—very simple ones. I open my morning newspaper. It tells me that if I would like to get a certain type of information, I must punch a certain number on my phone. If I do so, an automated voice comes on and tells me what my options are for further punchings. Ultimately, if I am lucky, I get what I want from the electronic voice. The mathematical underlay of this piece of informational technology is substantial. In other cases, where the information sought is not so well "pin-point-able," I may go six layers deep, end up frustrated, and curse the mathematical mind at work.

On the next page of the newspaper, there is a long article saying that computer identification of fingerprints has now been achieved. The article goes on to say that "millions of fingerprints have now been reduced to a mathematical formula."

Or take the weather map. That has been created automatically by graphical contouring routines that employ mathematical strategies. The business pages of the daily newspaper with their averages, indices, indicators, etc., are perhaps the most digitized, mathematized section of all, reflecting events on the market, many of which are pure conceptual potentialities of a mathematical nature and have hardly any immediate reference to the production, distribution, and consumption of food, clothing, or shelter.

Yet, despite the plethora of mathematics in which we are immersed, there is a general indifference, ignorance, even animosity, of the public to mathematics itself. Most often the mathematics lies hidden within various techno-logical structures (such as computer chips) and remains out of sight. This invis-ibility augments the lack of awareness, and contributes to an ignorance of the fact that the mathematical spirit and mathematical applications are not always benign.

And then there is the problem of communication. Mathematics is, in part, a specialized language. How shall we arrive at the goals outlined above with-out going into details that would be comprehensible only to people with deep training in pure and applied mathematics?

I conjecture, Dear Ghost, that when you were a student at William and Mary, you may have studied a bit of calculus on your own. This represents more mathematics than today's average American high school and possibly college student has studied. What I hope for in education today, given our limitations of time, space and language, is that we might achieve a level of understanding expressed in your letter to Adams: "though the details are not for every reader, the results are readily enough understood."

Goodbye for the time being!

Leibniz's Dream:
Mathematics and the Law

Gottfried Wilhelm Leibniz (1646–1716), the German philosopher and mathematician, dreamed of a universal language that he called the *characteristica universalis*. This was not to be a language for common usage such as Esperanto, but a language based on logic and mathematics in which distinct concepts were given distinct and precise symbols. By means of this symbolism and its rules of combination and manipulation, all human problems, whether of science, law, or politics, could be worked out rationally and systematically by means of computation. Here is the famous paragraph in which Leibniz expressed his dream:

> If we had it [the *characteristica universalis*], we should be able to reason in metaphysics and morals in much the same way as in geometry and analysis... If controversies were to arise, there would be no more need of disputation between philosophers than between two accountants. For it would suffice to take their pencils in their hands, sit down at their slates, and say to each other (with a friend as witness, if they liked): Let us calculate.

If, in today's world a conflict arose within the former Yugoslavia or between the Israelis and the Arabs, then instead of having a summit conference in some neutral city, one would simply gather together all the relevant facts, express them in the universal language, feed them into a computer, and push the button. All would then be hunky-dory.

Let me limit myself to one corner of Leibniz's vision: the interaction between mathematics and the law. My interest in this topic is that of a spectator, and I was led to it many years ago by my wife's cousin Michael Oakes Finkelstein, who has written extensively on the topic.

As a young lawyer working for the U.S. Government, Michael was confronted with the perception that juries consisted predominantly of whites even in areas where there were substantial number of blacks on the voting lists. Michael applied statistical methods and wrote about his findings. He went on to plunge deeply into the often turbulent and murky waters of law and mathematics, collaborating with expert mathematicians and statisticians such as Herbert Robbins.

I have to laugh when I think of how juries are now selected. At random? Forget it. There are lawyers and legal companies who specialize in jury selection (read: juror exclusion). As Cousin H. used to say, "Inside knowledge beats the stats ten to one."

Nonetheless, there is every reason to believe that mathematics will pervade the law more and more. I can imagine the following Leibnizian scene in the year 2037.

The Supreme Court of the United States is in session. No "Oyez, oyez." No longer is there an elevated bench with nine robed justices sitting in a row. No longer do elegant and eloquent attorneys present their cases orally and respond to questions from the bench. With the public recognition in the year 2020 that oral presentations were simply a part of an outdated ritual and not even cost-efficient, the courtroom was abandoned to the Smithsonian Institute, which maintains it as a museum, much like Independence Hall in Philadelphia.

The justices sit each in his or her chambers with several computer terminals. Each justice works with three law clerks who are computer specialists skilled at COMPLEG and other databases of laws, precedents, and decisions. Each justice has his own decision algorithms and programs supplied by the Softleg Corp. or another commercial spinoff from the country's law schools. The Softleg Corp. is now a subsidiary of what started out years ago as Metro-Goldwyn-Mayer. There are many such programs available, all tailored variously to the specific background, experiences, and legal philosophy of the individual justice. The preferred balance between justice and equity can be entered in percentages.

The constitutionality of the algorithms built into commercially available programs has already been argued before the court; its decisions were arrived at on the basis of previously existing legal software. (It was pointed out in argument that the whole of life has been built up slowly on millions upon millions of bootstrap operations.)

All relevant and preliminary facts, arguments, testimonies, and amicus statements are entered on line having been shipped out from the lawyers' offices. Each justice pushes a button and COMPLEG or its clones go to work to output a decision. If a justice is dissatisfied with the output, he or she may put in a call for more input, and this procedure is iterated until each justice is satisfied.

In a second available mode of operation, known as "inverse-law," the justice informs the computer how he wants the case decided and COMPLEG goes to work and produces a conforming opinion. So as to maintain a high degree of privacy, the law clerks have been dismissed halfway into this process.

Then the venue changes. The assembled justices meet in the flesh in the leak-proof decision chamber. Their personal terminals intercommunicate and produce majority and minority opinions from the collection of opinions of the individual justices.

Next comes the most solemn moment in the process: a ritual that maintains symbolically the continuity of four thousand years of legal practice. Each justice takes a slip of stiff cardboard and writes down a one (for the plaintiff) or a zero (for the defendant). These slips (called the Ostraca) are inserted into a fishbowl. An associate justice is selected as counter. The chief justice withdraws the slips from the fishbowl slowly one by one and calls out solemnly "zero" or "one" as the case may be. The counter marks these digits down, adds them up without recourse to a computer, and announces the decision.

Once the decision has been arrived at, the majority and the minority opinions are automatically collated and entered into appropriate databases. They are then available to the media and for use in subsequent cases.

The Peirces Take On Hetty Green:
Evidence via Statistics

I must have more than a bit of reporter's blood in me, because when I'm working on a piece of writing, I pump everyone around for information. E-mail makes it particularly easy, and when the pumpee is a relative who summers on nearby Cape Cod, it can make the job very pleasant.

When I told Michael about my "Jefferson" project with Bernhelm, and told him, moreover, that Jefferson knew a bit of mathematics and loved it, he immediately jumped in with, "Jefferson was a lawyer. Tell his ghost about statistics and the law."

"Great. Where shall I start?"

"Well, social statistics go back to the mid-1600s, when morbidity statistics were collected in London during the plague. But in the modern sense, they start with Quételet, after Jefferson's day, around 1830. But I have a better idea, why don't you start in the United States with the Hetty Green and Charles Sanders Peirce business?"

I remembered that when I was a boy the Sunday supplements used to run pieces about Hetty Green.

"Is the story scandalous?"

"Absolutely."

"Good. Tell me about it, then."

Michael told me about it and gave me a reference.

In the mid-1860s, Benjamin Peirce of the Harvard faculty and his son Charles Sanders Peirce, both mathematicians of the first rank, were involved as expert witnesses in an accusation of a forged will. The presumed forger was Hetty Robinson, the plaintiff in the case. Hetty was later to become the notorious Hetty Green (1835–1916), millionairess, recluse, litigious miser and a woman whom the newspapers of the day loved to call "The Witch of Wall Street." Two wills of Hetty's aunt Sylvia Howland were involved, a later will, leaving half of a considerable estate to her niece, and an earlier will leaving the whole estate to Hetty and containing an instruction that all later wills were to be ignored.

It seemed to the executor of the Howland estate that Hetty had forged the first will. He believed that two of the three signatures on it were too good and that they had been traced from a third signature. This, he claimed, invalidated the will. Hetty sued the executor.

The executor, to establish his position, engaged the Peirces. The Peirces analyzed the number of coincidences in position and size when upwards of forty pairs of signatures were superimposed. They concluded that the third signature, at least, had been forged. Their testimony supported the executor, but in the end, the decision in favor of him was reached in 1868 not through mathematics, but by a technicality in the Massachusetts law regarding court testimony about wills.

This was probably the first case in American legal history in which a probabilistic analysis was presented in evidence. (See a very interesting analysis of the Peirces' work by Meier and Zabell, *Journal of the American Statistical Association*, vol. 75, Sept. 1980, p. 497.)

One, Two, Three, Count it Legally

In the United States, there is a small but increasing amount of evidence brought forth via mathematical analyses. This methodological option has been turned around: the courts seem to have been asked increasingly to decide what is or is not appropriate applied mathematics in certain situations.

This point can be illustrated by alluding to a case of current interest that involves one of the most elementary of the mathematical operations: counting. Beyond a few score of apples or such things, counting as a physical act of ticking off one, two, three, etc. things is impossible. If we say that our bank balance is $643.18, no one assumes that we have sat down with a pile of pennies and counted them out in miserly fashion. And certainly when we say that the area of the United States is 3,615,208 square miles, there is no supposition that someone has marked out the terrain and counted it, square mile by square mile.

Counting, in the extended sense in which it is generally interpreted today, is an operation that may involve mathematical sampling theories of complex subtlety. Furthermore, the assumed existence of an objective count is a mathematical idealization. Counting in practice as opposed to counting in theory can be surrounded by vast ambiguities. The numbers accepted as the result of a "count" are social, economic, physical, mathematical constructs.

On Tuesday evening, January 19, 1999, William Jefferson Clinton, President of the United States, having been impeached by the House of Representatives and threatened with removal from office by the Senate, performed his constitutional duty of presenting the annual State of the Union Address. In this address, he said that he hoped that the the constitutionally required Decennial Census would be carried out according to the latest scientific methods. What was behind this remark on such a dramatic occasion?

In Article I, Section 2 of the U.S. Constitution, a section that deals with the House of Representatives, we read: "The actual Enumeration (of the population)...shall be made in such a Manner as they shall by Law direct."

Notice the ambiguities in this sentence. What does "actual" mean? What does "enumeration" mean? Who is to be counted? Only the people who actually answer the doorbell when the census taker rings or may they speak for others? What about people who are out of the country for a few months and have sublet their apartment to a non-citizen? What about illegal immigrants? What about the already conceived but as yet unborn? Should they be counted? Such rights have been asserted in other connections. How is the count to be made?

Recall that when the US Constitution was written and ratified by the individual states, a sizable fraction of the population was held as slaves and another fraction were indentured servants. Section 2 gives these categories special treatment as regards the number of representatives each state receives. Recall also that suffrage at that time was granted only to free men (not women!) of a certain age and of certain moral and financial qualifications.

At the time of the adoption of the US Constitution, Malthus was in his twenties. Adolphe Quételet, the Belgian mathematician turned social statistician, had yet to be born. The PES (post-enumeration survey), whose results can now can be placed in evidence in court, was far in the distant future; the variance-covariance matrices and their inverses, on which interpretations of the PES may be based, had not been dreamed of.

Now what is riding on the results of the census? First, the number of representatives each state may send to the House of Representatives. Second, the distribution of federal money to as many as 39,000 state and local governmental entities. Then, of course, since nosy census takers do much more than merely count noses, the decennial census becomes a tremendous database that all interested investigators may interrogate. In this way its content may influence future economic and social policy.

So, at the bottom line: census figures were heavily involved in the distribution of power and money. Considerations of these two—perhaps more of the latter than the former—are what led to *The City of New York et al vs. The U.S. Department of Commerce et al.* (88 cv 3474, argued in Federal District Court before Judge John McLaughlin).

The background of the suit was this: the City of New York claimed that the "raw" counts made by the Census Bureau (a division of the US Department of Commerce) seriously underestimated the number of Afro-Americans in New

York and that it was "arbitrary and capricious" for the Department of Commerce not to have ordered that the population figures be adjusted statistically to account for the uncounted. (It is an interesting piece of history that Thomas Jefferson thought that the *very first* U.S. Census underestimated the total "population.")

Eight statisticians, all of them well known, testified for the City of New York. Sixteen statisticians, all of them well known, testified for the Dept. of Commerce. (Thought: does the appropriateness of a mathematical method depend on the number of people who deem it correct?)

Arguing about the treatment of the previous (1980) census as a response to a number of suits instituted against the Census Bureau at that time, Eugene P. Ericksen, Joseph B. Kadane, and John W. Tukey wrote:

> We argue forcefully for an adjustment for we are convinced that it would improve the census. In our view the argument against adjustment is nonstatistical..We believe that the Census Bureau creates political difficulties for itself when it ignores the undercount. The Bureau will put itself in a better position by making its best effort, using available statistical and demographic methods, to adjust for the undercount. Errors will remain, but they will be smaller and we will no longer know in advance who is losing money and power because of the undercounting. (Ericksen, Kadane and Tukey, "Adjusting the 1980 Census of Population and Housing." *Journal of the American Statistical Assosciation*, Dec. 1989.)

Testifying for the Department of Commerce, statistician David A. Freedman said

> I would advise against adjustment. It seems to me there is no strong evidence on the table to show that adjustment will improve the distributional accuracy for states or other areas, and I think there is a real risk that adjustment would actually put in more error than it took out. (David A. Freedman, "Adjusting the Census of 1990," Tech. Rep. 385, Department of Statistics, University of California, Berkeley, April 1993. Also published in *Jurimetrics*.)

A counter argument to this, provided by Franklin Fisher, pointed out that while adjustment may not yield a better result, the errors will be random and that would be better than systematic errors.

There's an old proverb that when elephants fight, it's best to stay out of the way. So what is the poor judge, for whom such things as variance-covariance matrices or lost function analyses were not absorbed with mother's milk, to do when the experts disagree? And what is the legal position of "expert" testimony?

The *New York Times*, on Sunday, July 4, 1993, reporting Supreme Court decision No. 92–102, *Daubert vs. Merrell Dow*, indicated that Justice Blackmun said for the court, that while the actual conclusions of expert witnesses need not be "generally accepted" in the scientific community, the "methods and the procedures" used in reaching those results must be valid. The judge should limit the jury's consideration to testimony or evidence that is not only "relevant, but reliable."

Reacting to the *Times*' report and not inquiring further into the guts of the full decision, I should be inclined to say: this is all fine and dandy, but how does one decide who is qualified to say what is relevant and what is valid or reliable in matters of specialists' concerns when the specialists themselves cannot agree?

To return to the census dispute. As regards the 1980 census, the court found that no adjustment was required. As regards the 1990 census, the court again found that no adjustment was required. But the judge added that the decision was appropriate to a case formulated on the plea of an "arbitrary and capricious" refusal to adjust. If the case were tried *de novo*, he would find for adjustment.

On January 13, 1998, the head of the Census Bureau resigned. She was in favor of adjustment by sampling. The thought is that sampling will increase the power of the minorities and is correspondingly a political rather than a mathematical issue.

The issue was argued before the US Supreme Court on November 30, 1998. The President's State of the Union plea was rejected. On January 25, 1999, the court decided 5–4 against sampling in the Year 2000 census to determine the number of representatives in each state. But the Court seems to have left open the possibiity of sampling for other purposes, for example, individual states may use adjusted figures to determine their own state representatives, or for purposes of redistricting. The Census Bureau said it would issue two sets of numbers.

The act of counting is context sensitive? An old story in the business world where multiple balance sheets are often prepared. But multiple answers for the population? What answers to use in what instances? To what litigation and chaos will this ambiguity lead?

As mathematizations enter and format our lives more and more, not only in the technological sphere of life, but in the socioeconomic sphere, the number of instances of disagreements of the "experts" has multiplied. There is every reason to think that this tendency will continue. The position of mathematical evidence in the courts is still moot though there has been a fair amount of practice. (See Michael O. Finkelstein and Bruce Levin: *Statistics for Lawyers*, 1989, for discussions of many specific cases that have come to trial. Also: *Quantitative Methods in Law: Studies in the Application of Mathematical Probability and Statistics to Legal Problems*, 1978.)

Even when experts agree, and this does happen, friends who have testified as expert witnesses have told me that a clever lawyer, brought up in the adversarial system and using the well-developed rhetorical ploys of this system can, in microseconds, reduce their testimony to hash in jurors' minds. And then there are the interested constituencies: in the census case, the local politicians, the social workers, the applied statisticians as a professional group, special interest groups for the nebulous uncounted, etc., all of whom might submit *amicus curiae* briefs. The courts are not really compelled to descend (or ascend) to mathematics to reach their decisions. And so we are back to the experiences of the Peirces, father and son.

Tartaglia's Dilemma:
Mathematics and War

The lectures I gave in the spring of 1992 at the University of Roskilde may have disappointed the more dedicated firebrands who came to hear me; I raised more questions than I answered and I presented no real agenda of action. The typical disease of thinkers? Remember Marx: the problem is not to analyze the world but to change it. But then again, Rodin's *Thinker* is not depicted on the ramparts the way Delacroix's revolutionaries are.

I believe that the mathematical spirit both solves problems and creates other problems. What is the mathematical spirit? It is the spirit of abstraction, of objectification, of generalization, of rational or "logical" deduction, of universal quantification, of computational recipes. It claims universality and indubitability. I have the conviction, shared by Bernhelm, that this spirit is now (unlike in Jefferson's day) pushing us too hard, pushing us to the edge of dehumanization.

This is hardly an original perception. As early as 1920, the Russian author Yevgeny Zamyatin described a highly mathematized social distopia in his sci-fi novel *We*, a book that is a precursor of both Aldous Huxley's *Brave New World* and George Orwell's *1984*. Now the world and all that is therein are increasingly chipified.

The successes that mathematics has enjoyed has been incredibly—often incomprehensibly—great. Quite naturally, its failures, which are many, have been hidden in the closet.

Consider as an example modern warfare. Through advanced science and technology, warfare utilizes many mathematical ideas and techniques. The creation of vast numbers of new mathematical theories over the past fifty years was due in a considerable measure to the pressures and the financial support of the military. The utility of mathematics in warfare was already recognized in the days of Archimedes (200 BC) and the moral dilemma was put forth already in the sixteenth century by the mathematician Niccolo Fontana (Tartaglia) (c.1500–1557).

Tartaglia ("the Stammerer"; the result of a facial wound received as a child during the siege of Brescia, his home town) is known in the history of mathematics for his connection with the solution of the cubic equation and for the subsequent controversy over priority and publication rights with the better known mathematician Geronimo Cardano (1501–1576). What is not mentioned in the two large histories of mathematics I keep on my desk and what I found out quite by accident only ten years ago, is that Tartaglia wrote a book on ballistics, La Nova Scientia (1537), that contained a remarkable preface. Histories of mathematics, particularly the shorter ones, tend not to deal much with the applications of mathematics, preferring the purity of the unapplied; in any case they shy away from the embarrassing fact that over the centuries, in one way or another, one of the principal applications of mathematics has been to military matters.

The Lownes collection of rare scientific books in the John Hay Library of Brown University has a copy of Nova Scientia, and I was able to examine it. The frontispiece of the book is an allegorical picture that places ballistics (the New Science) firmly within the context of Renaissance humanism.

A tower of wisdom and knowledge is depicted consisting of three cylindrical walls of diminishing radii stacked concentrically to form higher levels. Each wall surrounds an open area. At ground level, the figure of Euclid stands at a door ready to guide men up into the tower. On the first level, two cannons with different angles of elevation have been shot off by their attendants. The different trajectories of their cannon balls have been traced out. Nearby stands the author, Niccolo Tartaglia, surrounded by a variety of female figures labeled Astronomy, Music, Arithmetic, Geometry, Poetry, and Astrology. At the second level stand the figures of Plato and Aristotle; and at the uppermost level the female figure of Philosophy.

What we have here is the hierarchy of knowledge as set out by St. Thomas Aquinas, and about which I first learned from the lectures of Philipp Frank. But it's a tower that lacks its topmost thomist level: theology!

At the bottom of the allegorical frontispiece are written the words:

> Disciplinae mathematicae loquuntur <vobis>
> qui cupitis rerum varias cognoscere causas.
> Discite nos cuncta haec per unam viam.

("The mathematical disciplines speak to you who desire to know the various causes of things. Teach us all these things in one way." Translation courtesy of David Pingree.)

The book discusses the trajectories of cannonballs. The theory is pre-Galilean and a solution is arrived at in terms of straight line segments and arcs. That does not concern me here. The preface of the book does. There Tartaglia speaks of a moral dilemma. He wrote:

> Through these discoveries, I was going to give rules for the art of the bombardier... But then one day I fell to thinking it is a blameworthy thing to be condemned—cruel and deserving of no small punishment by God—to study and improve such a damnable exercise, destroyer of the human species, and especially of Christians in their continual wars. For which reasons, O excellent Duke, not only did I wholly put off the study of such matters, but also I destroyed and burned all my calculations and writings that bore on this subject. I much regretted and blushed over the time I had spent on this, and those details that remained in my memory (against my will) I wished never to reveal in writing to anyone, either in friendship or in profit (even though it has been requested by many), because such teaching seemed to me to mean disaster and great wrong. But now seeing that the wolf [i.e., the Turkish Emperor Suleiman] is anxious to ravage our flock, while all our shepherds hasten to the defense, it no longer appears permissible to keep these things hidden.

It is hardly possible today to put the dilemma more clearly: the knowledge that mathematics is of use in warfare, the secrecy (for whatever reason) attending such use, the guilt that a humane person might feel in engaging in such work, the assuaging of the guilt by the realization that the survival of one's country or one's own survival may be at stake. The dilemma is not yet resolved and cannot be, short of eliminating war; and it is deepened by the ambiguity borne out over and over again that out of good can come evil and out of evil can come good. A number of the technological innovations of the past half century, things we would be loathe to give up, can be identified as spin-offs from World War II. These things, logically speaking, technologically speaking, did not require war to bring them forth, but the historical case is plain to read.

A little ancient history will help us position the relationship between mathematics and war a bit more accurately. While Tartaglia may have been the first to raise some moral or strategic doubts, he was not the first to talk about the connection. The ancient Babylonians made "siege computations," e.g., what volume of earth, how many bricks, etc., for a siege ramp. A description can be found in Otto Neugebauer's 1933 work *Babylonische Belagerungsrechnung...* (*Babylonian Military Siege Computations.*) Socrates says that a military commander needs to know some arithmetic and geometry for arranging his troops optimally. (Plato, *Republic*, 525b.) Hero of Alexandria in his *Metrica* and *Dioptra* talks about the use of optical equipment for surveying terrain that is occupied by an enemy.

The Arabic mathematics that preserved and transmitted the classical Greek corpus to early modern Europe was only occasionally concerned with military matters. Contemporary middle eastern military leaders profit from the fruits of modern mathematics.

In the Renaissance, the study of mathematics engaged the practical world in architecture, perspective painting, artillery and ballistics, cartography, and commercial arithmetic. The fourteenth-century Portuguese court of Henry the Navigator supported the development of navigational mathematics and maintained a policy of strict secrecy with respect to it. In this way, the applications of mathematics advanced Portuguese commerce, naval warfare, and government sanctioned piracy (letters of marque). Navigation was one of the driving forces behind new representational and computational techniques, including cartographic principles and the invention of logarithms.

In the seventeenth and eighteenth centuries, various scientific academies were founded with royal sanction and offered prizes for the solution of practical mathematical problems. Since the strength of the state was often defined in military terms, the academies became the interpreters and transmitters of military mathematical needs. Goal directed research was common. "Hooke's Law" (Robert Hooke: 1635-1703) in the science of the strength of materials was, in all likelihood, a theoretical spin-off from empirical inquiries on the elasticity of wood. The Royal Navy wanted to reduce its consumption of wood in shipbuilding.

Mathematics was an integral part of the training of ships' officers as well as military officers. Fortification mathematics and ballistics based on the Galilean parabolic trajectory were taught. However, it seems that in practice, the mathematics was reduced to rote and no new mathematical developments sprang from the officers' schools.

The French Revolution ushered in a new phase in the relation between science and technology. The *Ecole Polytechnique* was founded in 1794, and within ten years it was made part of the ministry of war! During these years, the training it offered was limited to mathematical and semi-mathematical subjects: pure analysis, applications of analysis to geometry, mechanics, descriptive geometry, and drawing.

While Diderot's *Encyclopaedia* declared that "war is a fruit of men's depravity; it is a convulsive and violent illness in the body politic," by the time of the Revolution and the Napoleonic wars, there had been an attempt to reconstruct life along scientific lines (e.g., a new calendar, uniformization and metrization of weights and measures) and hence to reconstruct war along such lines. The larger, more powerful countries began, slowly, to organize themselves for total war.

Although a close association between mathematics and the military was inaugurated at the *Ecole Polytechnique*, nothing much seems to have come of this association in its early years. One possible exception is the case of Gaspard Monge (1746–1818). While teaching at the *Ecole Militaire de Mézières*, Monge developed his theory of descriptive geometry, probably as a result of his experience with fortification mathematics. Later, as politician, minister of the navy, and as founder and teacher in the *Ecole Polytechnique*, he was instrumental in changing the mathematical curriculum. The influence of the French mathematicians of the Revolution and the Napoleonic era was tremendous; their work affected mathematics and its later applications both to civilian and to military purposes. For example, the ideas put in place by Fourier today constitute the basic ingredients of theoretical information transmission and processing.

The applications of science and technology to war, as a concerted effort, began with World War I. The most important areas were chemistry and metallurgy. Physics was less important, and mathematics, if one excludes exterior ballistics (trajectories of artillery shells), hardly at all. On the other hand, airplanes were then ten years old, and the science of aerodynamics (which is highly mathematized) got a lift from the use of planes as fighters and reconaissance. In the United States, the NACA (National Advisory Committee for Aeronautics, precursor of NASA), a highly mathematized research institution, was established by Congress shortly after World War I.

Reviewing World War II from the point of view of scientific and technological developments, one thinks first and foremost of the atomic bomb, then, perhaps, radar. A full list would include developments in aero- and hydrodynamics, ballistics, aeroballistics (air-to-air), rocketry, guidance, gun control, analog and digital computers, cryptography, underwater sound detection, mine clearance, salvo efficiency evaluations, sequential analysis leading to quality control, strategic analyses and simulations, linear programming, optimization theory, and operations research in general.

In World War II, many branches of advanced mathematics were found to be applicable to the technologies just mentioned. In most instances, the mathematics pre-existed, and was simply used "off the rack," as it were. Most developments of military applications were simply deepenings of standard theories. In some few instances, one might cite operations research and the theory of prediction and control, genuinely new material emerged that later became autonomous fields.

In the half-century following World War II, fifty years of the cold war confrontation between the US and the USSR, and despite cautions sounded by President (formerly, General) Dwight Eisenhower, the military-industrial complex flourished and was thoroughly mathematized. This was so much the case that in the early days of the Cold War, one heard it said that while World War I was the chemist's war and World War II was the physicist's war, World War III, if it ever came, would be the mathematician's war.

Through its system of contracts and grants to both universities and industries, the army, navy and airforce fostered the growth of both pure and applied mathematics and statistics. The number of colleges having mathematical programs for advanced degrees grew, as did the number of academic and industrial mathematicians. In the US, applied mathematics, which had not been a favorite of the pre-World War II mathematical establishment, was given new importance and intellectual stature.

Numerous new professional societies having mathematical interests were established; e.g., SIAM (Society for Industrial and Applied Mathematics), ACM (Association for Computing Machinery), Operations Research Society. The number of professional mathematical journals grew mightily. The number of national and international meetings increased, as did the number of national meetings dedicated to very special mathematical topics and techniques.

Given the full purses available during the Cold War, the universities easily adjusted themselves to the generous overheads allowed by the contractors, and to the numerous fellowships offered to graduate and post-graduate students. The granting agencies did not examine too closely the military utility or potential of the investigations they supported financially, and much non-military mathematical work was accomplished as a result of this benign neglect. A claim for support was often put forward on the basis that one can never tell in advance when a piece of pure mathematics will become applicable and the history of mathematics displays many instances of the truth of this assertion.

One thing is perfectly clear: the science of mathematics in all its aspects, pure and applied, national and international, benefited greatly from the recent wars, hot and cold.

They Shall Beat Their Swords into Fun and Games

I never imagined that a visit to DeepPlum, Inc. (I have used a pseudonym) in Silicon Valley, California, would engender thoughts about the old problem of the relationship of form to content. Allow me to get there in slow steps.

In January, 1997, having a family affair in San Francisco to attend, I topped off my trip with a one-day visit to DeepPlum courtesy of Alison Jayne, a former Ph.D. student who was employed there. DeepPlum concentrated on computer products (hard and soft) for graphical purposes, but now, especially since it has reorganized itself upwards, it offers a wide variety of products.

Spread over more than twenty buildings, with fountains splashing here and there, with new buildings coming up in bright decorator colors, with reception rooms furnished like conversation pits, with tens of thousands of employees in the US and abroad, the DeepPlum campus shouts out to the visitor "success, activity, cutting-edge" in no uncertain terms. The self-image of a company is often expressed in the architecture it commissions at the height of its prestige. I thought of Pennsylvania Station in New York City (now sadly degrandiosed) or the stunning General Motors Research Labs designed in the mid-Fifties by Alvar Aalto.

The interior of the DP buildings are quadrangularized into warrens in which Ph.D.'s in computer science and/or mathematics abound as plentifully as dandelions in May. What are they doing, these Ph.D.'s? I have only two data points to share with my readers. A recent arrival from one of the leading graduate computer science departments told me about his thesis. In computer animation, if you want, say, to move a human arm from an initial position to a final position, it won't do to interpolate linearly in time. The movement will appear inhuman. You must accelerate and decelerate the arm. My informant's scheme was to move the arm in such a way that the energy of the motion would be minimized. He would be delighted to give me a demo.

He logged on. A tennis player was brought up on the screen and the player's movements leading to a perfect return shot looked good to me. I asked the new arrival if that is what he was working on for DeepPlum and he replied, "No. I'm working on triangulations for auto surface rendering." I assumed he was adapting to local usage some of the dozens of surface triangulation schemes that now reside in textbooks.

A second demo in another building had been arranged with another member of the technical staff. He flew me over a portion of the California-Nevada topography that had been pieced together triangularly from satellite shots. We soared at will. We zoomed up and down.

"Zoom in on Silicon Valley."

He zoomed in.

"Perhaps we can zoom in on ourselves?"

"No," he answered with some seriousness, "DeepPlum hasn't yet solved all the problems of self-reference."

Mathematics, I recalled, abounds in self-references; all the way from $x = 2x - 1$ to Russell's Paradox.

What I saw was impressive, but I wondered whether I was being served up a few refried beans, so to speak, the real new and hot stuff being company confidential and reserved for potential customers. I had heard that mathematicians in some companies have been irritated by the surveillance of company lawyers who think that the answer to two plus two should be kept under wraps.

During a gap in my timetable, I asked what DeepPlum had cooked up for its Webpage. Click, click and we were (rather by accident than by intent) into the file of the vitas of DeepPlum employees.

"Ask the Web what the average DeepPlum employee age is."

My host asked and answer came there none.

"I know it anyway. It's 32 years."

"They're all bright people, obviously, but their lips are still wet with milk. And will DeepPlum love them when they're 64?" I asked, paraphrasing the only Beatle song I knew.

"Well, I'm on the shady side of 32."

"Ask the Web what the mean employee salary is. No, ask for the median salary. That would be a more significant statistic in view of top brass superpay."

The Web was queried, and answer came there none. I was delighted that the Web treated certain personal parameters with respectful silence. Control of information is important; the existence of cyberinfo merely strengthens this home truth.

The time had now come when I was to visit and interview the senior vice president for research. En route to his conversation pit, I accidentally bumped into the CEO and chairman of the board (since bounced), the only man I met at DeepPlum who was wearing a suit. He welcomed me to "the campus."

From the senior VP for research, I learned that their total revenue in the previous year was several billion dollars: evidence of good machines, good software. I learned that DeepPlum was international.

I told the senior VP I had heard of the use of skillful third world programmers who work at one-tenth the salary of their American counterparts. The VP simply reiterated, "Well, we're an international firm. Of course we hire locals."

I asked what mathematics curriculum he would recommend for future DP employees, and I was surprised when he answered that the standard old-fashioned core curriculum (calculus, elementary and advanced, linear algebra, probability, etc.) was quite adequate. Was he being disingenuous or must the cutting-edge stuff always be learned in-house?

The senior VP was a rather quiet, circumspect man who never pushed the conversation one sentence beyond what was necessary to answer the question put. I tried one more gambit.

For some months I had been on the mailing list of a very fancy in-house magazine called *Iris* put out by Silicon Graphics, a competitor of DeepPlum. In October, 1995 a lead editorial in *Iris* opined that "In the past, technology (in the US) was driven by defense. Today it's driven by entertainment." I asked whether the thinking at DeepPlum would go along with this opinion. The VP skirted the question and in so doing, gave me what I considered a stimulating answer.

"Here at DP, we don't do 'content creation.' That's done by our customers. We provide the frames into which the content can be poured and processed. Our developments are driven by the needs of the 'content suppliers.'"

The interview was over. I wondered whether the questions I asked and considered innocent enough had intruded into corporate privacy.

On the flight back to Providence, I meditated on the last answer the VP gave me. "Content creation" and "content suppliers" were new phrases to me. Perhaps this was merely commercial jargon intended to convey that DeepPlum's products must not be too narrowly focused. But I interpreted the phrases literally.

I recalled that critics and historians of literature, poetry, art, occasionally made a distinction between form and content. By this token, DeepPlum was a manufacturer of forms and not of content. I recalled that Aristotle, imagining a sculptor who has the idea of a figure in mind before he commits it to clay, said that form determined content. Two and a half millennia later, Marshall McLuhan agreed with him and coined the catch phrase "The medium is (or becomes) the message." The man I interviewed asserted the reverse: content determines form: we produce a content-free product. And yet I thought that the demos I was shown, the tennis player and the topography of California, were content-laden. Can form be exhibited independently of content?

But what is form and what is content, and can they really be distinguished? Is the bagel a form and its ingredients the content? Is the shotgun a form, and the dead duck the content? Is the layout diagram of a chip content or is it form? Is $a^2 + b^2 = c^2$ form or content? Or do all these instances have to be contextualized before a decision can be made?

A century ago the literary critic A. C. Bradley said that form and content are inextricable. "If substance and form mean anything in the poem, then each is involved in the other and the question in which of them the value lies has no sense." ("Poetry for Poetry's Sake," in *Lectures on Poetry*, 1909.) Werner Heisenberg (1901–1976), Nobelist in Physics, 1932, wrote

> Mathematics is the form in which we express our understanding of nature but it is not the content of that understanding. Modern science and, I think, also the modern development of art are misunderstood at a crucial point if we overestimate the significance in them of the formal element. (*Across the Frontiers*, p. 146.)

Enjoyable though they were, what was memorable about my visit to DeepPlum was less the demos than something else. The avowed separation of form and content enhanced my appreciation of the extent to which our manufacturing civilization is profit-driven, whether through defense or through entertainment. (So what else is new, sonny boy?) Whether through industrial globalizations or through the HMOs at home, we are no longer emotionally in the business of producing products; we are in the business of making money. In their pursuit of this bottom line, in their production of hard material objects or of soft conceptual objects, companies are forced into a driven mode that splits the creator from the creation, the form from the substance. This mode has demonstrated its power many times over, but is only now, in the consciousness of the public, beginning to demonstrate its dark side.

Mickey Flies the Stealth

One aspect of the split between form and content showed up in a report that came out of Washington: *Modeling and Simulation: Linking Entertainment and Defense*, (The National Research Council, National Academy Press, 1997). Reports that come out of Washington tend to be deadly dull stuff, but this one, based on a two-day workshop held in October 1996 at Irvine, California, caught my attention.

The purpose of the workshop was to discuss problems common to the entertainment industry and the defense industry vis-a-vis modeling and simulation. More than fifty people, some from the film, video-game, location-based entertainment and theme park industries, some from the DOD, defense contractors, some from universities, met face to face and tried to decide what interests they had in common and how "two communities that have tended to operate independently, developing their own end systems and supporting technologies," might help one another. I wish I had been present.

The principal goals of modeling and simulation discussed in the report are training, analysis, and the acquisition of systems with new capabilities. While the current products of the entertainment sector are known to the generality of mankind, the discussion also includes a projection of the next generation of video games:

> The intent of on-line gaming is to create massively networked games in which hundreds, if not thousands, of players, can play in the same virtual world simultaneously.... In the multiplayer mode, the player will enter a persistent universe that will run 24 hours a day, 365 days a year. Players will be able to join a game whenever

they want and in whatever role they want (tank commander, fighter pilot, etc.) They will be immersed in an environment of teammates and adversaries controlled by other players and the computer with the distinctions between the two becoming increasingly hard to detect.

(Next generation? Isn't the gigantic on-line stock market one vast game with millions of simultaneous players?)

What, more precisely, are the areas of overlap between entertainment and defense? According to the press release that accompanied the book, the principal ones are:

(1) Virtual reality technologies. The creation of immersive simulated environments against which defensive-offensive action takes place.

(2) Technologies and standards for networked simulator systems. Rapid communication between many (thousands) of simultaneous players.

(3) Computer generated characters (i.e., people). The challenge here is to develop characters that model human behavior in activities such as flying a fighter aircraft, driving a tank, or commanding a battalion such that participants cannot tell the difference between a human-controlled force and a computer-controlled force....The current limited responses of the antagonists require that additional intelligence be built into them. Smarter computer opponents could learn how different players operate and then adapt their responses accordingly.

In my role as a writer, I was intrigued by reading that part of the development of generating characters is that of "electronic storytelling." The objective here is to create the right set of stimulants, visual, aural, olfactory, vibrotactile, so as to elicit an efficient set of psycho-biological responses in a real battle arena.

Hollywood and TV craftsmen have successfully algorithmized numerous emotions such as fear (reported by W. D. Hillis of the Walt Disney Co.); think of walking slowly through a dark house into what seems to be an unoccupied room. The music flares up suddenly and the camera reveals...). Cybersex is so well advanced that a filtering V-chip is an object of freedom of speech controversy. Much more recently, the newspapers reported that via computer graphics Japanese cyberartists have created such debilitating nausea and convulsions that viewers (mainly children) have been sent to the hospital. All this gut stuff is very easily transferable to simulated military engagements: cyberwar.

Entertainment and war! One might think: what an odd combination; what strange bedfellows the computer has made. A moment's thought reveals the truth: it did not take the computer to link entertainment and war.

"Ever since words existed for fighting and playing, men have been wont to call war a game," wrote Johann Huizinga in his classic 1938 work, *Homo Ludens: A Study of the Play Element in Culture*. The two activities have been close to each other ever since primitive man poked a spear into the belly of his enemy and found that the outcome produced pride, loot, women, and a tremendous high. When knighthood flourished, troubadours sang of love and glory, while elaborate jousting contests, far from mock, provided entertainment for the ladies who sat in the stands wearing the colors of their favorites.

At the end of World War II, I attended a victory celebration given by the Air Force at its Langley Field, Virginia base. As I recall it, the main talk was given by the commanding officer of the base who spoke of military activity completely in the metaphor of football: "we gained thirty yards; we ran it for a touchdown," etc. Here also war is made acceptable emotionally by linking it with entertainment.

In the opposite direction, think of chess and go. Both very old, very mathematical games, are war games. In chess, when your opponent's king is taken out (modern terminology) the game is won. (The word checkmate is derived from the Arabic: shah-mat: the king is dead.)

The relationship between entertainment and war becomes vividly clear when contemporary entertainments such as football are pursued with the players dressed in full suits of heavy protective armor; when high school coaches are out for blood, and their young charges cry when they do not make the first team; when video games are based ninety-nine percent on the kinematic responses of the players to aggressive situations.

In 1910, when the Spanish-American War was just behind him, the famous American psychologist and philosopher William James wrote an often-cited essay that explored the possibility of a moral equivalent of war. He came to a somewhat pessimistic conclusion: that any moral equivalent of war would have to come to terms with the innate pugnacity of humans, with their love of glory, with their fascination with horror, their feeling that war is the strong life, with the patriotism that accompanies it, with the romance that results from it, and with the determination to exterminate foreigners for the greater glory of God.

On the other hand, James wrote, not entirely with disapproval,

> The apologists for war pose the alternative: a world of clerks and teachers of co-education and zoophily, of "consumer leagues" and "associated charities," industrialism unlimited and feminism unabashed; no scorn, no hardness, no valor. Fie upon such.

And to this list, he might have added: no ability to provide for the common defence. James concluded that

> martial values must be the enduring cement; intrepidity, contempt of softness, surrender of private interest, obedience to command must still remain the rock on which states are built.

And the only concrete moral equivalent that he offered was to institute a youth corps: draft young men to do a lot of hard and often unpleasant physical work; and to work not abroad (as with our present day Peace Corps), but at home.

Now bring in mathematics. The two-way relationship between mathematics and defense is well known and needs no elaboration here. So let me move on and talk about mathematics as entertainment. This notion is familiar to teachers who use the idea of "math as fun" as an inducement to learning. Math as fun is also asserted by researchers who argue for public funds without surveillance. "Let us play," they say to the grantsmen, "with no strings attached to our play, for one can never tell what goodies (in the public's sense) might emerge." Amongst themselves, privately, for the idea would be political poison, they say simply "let us play."

For the cognoscenti, the cat is really out of the bag. George Steiner, a literary critic of great reputation, a historian of ideas and a math buff, talks about mathematics as play in his introduction to my edition of Huinziga:

> Of all human activities, mathematics—particularly pure mathematics—comes closest to Huinziga's own standards of elevated play....What is more playful in the deep sense, than, say, the Banach-Tarski Theorem, whereby we may divide a sphere as large

as the sun in such a way that the whole may be fitted in our pocket?

In the reverse direction: entertainment as mathematics? Think of the total mathematization of sports in the past half-century; the stats that now determine strategies and salaries. The triangle is now closed; and it moves in both directions: mathematics-entertainment-war-mathematics.

Yes, the products in which entertainment and defense have been married can be employed for military training on ground, air or sea, or for the study of strategic alternatives. Or, on an entirely different level, the marriage can provide history buffs with the ability to sit in their programmers' chairs as Monday morning quarterbacks and reenact the Battle of Gettysburg, or the Peloponnesian War, or the capture of Passchendaele made realistic via the Fourier transform and made even more so by having random number generators and NURBS-animation strew them with virtual victims in their last agonies. The game becomes a military theater of Grand Guignol with audience participation.

The lines between real war and virtual war become more and more blurred. In the movie *Wag the Dog*, which some viewers find screamingly funny and others depressingly prophetic, the words "War is show business— that's why we're here" are spoken by a spin-master who has contrived a fake war to rescue a President from a sex scandal. The scenario certainly serves to blur even more within today's idiom the distinction between the real and the virtual.

But the convergence of the real and the virtual brings to my mind a more cheerful and wildly visionary thought: is it possible that in this convergence we may find the moral equivalent of war sought for by philosophers? Is it possible that future Agincourts or the recent siege of Sarajevo might be played without real casualties on the fields of networked simulator systems? The medieval world occasionally settled its disputes by the single combat of designated champions. Could a similar notion be projected into the new millennium via computer simulation? Would it suffice if the battle of the robots, material or virtual, the battle between my computer package and yours would be accepted by raging nations as determinative? If one grants the psychological necessities set out by William James, my optimism falls to the ground. I don't see that virtual warfare comes close to fulfilling his conditions.

If, as predicted by some Siliconites, the principal future product of the United States should become entertainment, the equivalence of entertainment and defense might prove as explosive a mixture in the social sense as the introduction of the gunpowder that spelled the end of the castle fortresses. Yet if simulated and networked military engagements were to result in the destruction of all the virtual players, some small measure of sense might thereby seep into the collective brain of humanity.

Personal Choices:
What Can an Individual Mathematician Do?

It is easy to read Voltaire's *Candide* as an allegory on modern science and technology. The three main characters are Candide, Master Pangloss, and Cunégonde. Candide is Everyman; open, fair, a bit naive and gullible. Master Pangloss proclaims: despite momentary setbacks we will reach the best of all possible worlds through technology. The beauteous Cunégonde, who at the end of the story has suffered greatly and has turned ugly physically, is the exterior world, raped, mistreated, thrown onto a cosmic dungheap.

I admit that this interpretation turns Voltaire's tale on its head, because Voltaire is pro-science. He describes the sane, felicitous life led in an El Dorado or Shangri-la-like kingdom, in which the king maintains a two hundred foot long laboratory for experiments. (Some Voltairians have conjectured that his El Dorado derived from his impressions of early eighteenth-century Quaker Pennsylvania!)

Given the inherent ambiguities of their position, what should a sensitive mathematician or scientist do in response to Tartaglia's Dilemma? Is it the gun that kills or the people who use them? Is it the ideas that are lethal or the people who implement them?

Before World War II, almost all the research mathematicians were employed by universities. With the coming of WW II, many left the university for military services, serving either in uniform or as civilians, or working in so-called war industries. Young scientists with bachelors' degrees were also recruited. As I have mentioned before, I myself served for three years as an aerodynamicist with the NACA. Some mathematicians served active military service assigned to duties that had a bit of mathematics associated with them, such as navigators on ships and planes.

During the period of the Cold War and the Vietnam War, the number of mathematicians who objected to the use of their subject for military purposes multiplied and their options for action varied. There were mathematicians who

• Left the profession entirely.

• Immigrated (or returned) to a country whose military involvement was small.

• Kept some distance from military applications in their work and teaching.

• Accepted military grants and rationalized their action by saying: "Well, the material I am working on has no chance at all of being applied, so I am doing good by taking away money from projects that might yield such applications." (Cf. George Bernard Shaw's *Major Barbara*)

• Worked on problems emerging from or aided by technology and seen as socially desirable: e.g., medicine, pollution control, oil consumption minimization, economics, etc.

• Worked politically and institutionally for scientific and moral responsibility.

• Involved themselves politically, both at the personal level and through association with pressure groups and at various levels of propagandistic rhetoric.

• Engaged in violent rhetoric, protest marches, strikes, or terrorism against mathematical and computer installations or against individuals, random or selected.

Thus people individually make peace with their consciences and with Tartaglia's dilemma. But for society as a whole, the problem remains deep and unresolvable. As long as war persists as a social option, science, technology, and mathematics will be associated with it.

Education:
Constant Reformation, Constant Complaint

There are nine and sixty ways of constructing tribal lays
And-every-single-one-of-them-is-right!
　　　　　 – Rudyard Kipling, "In the Neolithic Age."

Reminiscing about Mary Cartwright and the nine-point circle raised in my mind the perennial question: what and how and whom shall we teach? And to what end? Jefferson learned the mathematics of surveying. Today what with the new "total stations," virtually all the calculations are pre-programmed so that the surveying team needs only to sight known benchmarks and points whose position is to be determined. Practically no serious calculational drudgery is required to do the job; even contour maps are interpolated automatically.

When it comes to education, nothing is sacred really, except what people declare to be so. Under the doctrine of the separation of mathematics and theology, God may be declared out of bounds in formulas, but secular sanctities may also be questioned.

Take calculus, for example, that most beautiful structure of ideas created by Newton and Leibniz and augmented by the work of hundreds of others. Who needs it? Is it really important for a liberal college education?

For the past twenty-five years, instruction in calculus has been under attack for a variety of reasons. "Discrete mathematics" is said to be more relevant to the cutting edge of applications. The real number system, on which calculus is based, is infinitary in nature, and hence is simply one of the great myths of mathematics, beyond the range of digital computers. Despite these and other home truths, calculus is taught widely as the next higher mathematical subject after trigonometry and coordinate geometry. Calculus textbook writers have amassed fortunes.

In the years when the so-called New Math raised enthusiasms and applause combined with groans and condemnation, I advocated a manner of teaching elementary calculus that I called the "computer calculus." This was in the second generation of electronic computers (e.g., the IBM 650) when data was entered and taken out via punched cards. It was clear to me and to all who gave it a moment's thought in those pre-Maple, pre-Mathematica, pre-PC, pre-graphical laptop days, and pre-Webbed world, that programs could be written that would output automatically the answers to 90 or 95 percent of the plug-and-chug problems that were traditionally assigned in elementary calculus.

It was time, I thought, for calculus education to recognize this fact and to reform the manner in which the subject was taught. Now calculus education, ever since it emerged from the brains of Newton and Leibniz, has constantly been reformed; as a proof of this statement just open up *Principia Mathematica* and ask yourself whether you would want to teach calculus out of it. (Perhaps they do it in St. John's, a college that advocates a great books curriculum.)

I had some ideas as to how I would like to teach elementary calculus, but since that subject was not taught in my (applied) mathematics department but in the (pure) mathematics department up the street, I had to get their permission. I had no inkling of the storm that my request would create.

"No, no," they said (in principle), "what you're suggesting is absolutely wrong mathematically and totally inadequate. It is heretical. We will not allow the computer to sully the beautiful purity of the hypothetico-deductive methodology that can be exhibited within elementary calculus." I applied more pressure, and was able to break through with permission to teach experimentally one section of elementary computer calculus within the (pure) mathematics department.

As the beginning of the semester approached, I had to arrange for computer time for my class. I went to the mathematics department and suggested that as this was a course in their department, it would be nice if they went to the computer manager and requested a block of time for my students. A second storm arose. "No. No. No. We don't want to dirty our hands by making such a request." (These quotes should be interpreted spiritually, not literally.)

Well, the upshot of all this was that I got the computer time through my own department and I gave the course to about 25 students. The course had, inevitably, its successful parts and its failures. But I thought it was sufficiently successful so that the following year, I went around the country as an NSF Chautauqua Lecturer promoting "computer calculus."

I then bowed out of teaching elementary calculus, and the world of theory and technology went on to bring forth non-standard analysis (calculus), CD-Roms, and the Web without any help from me.

What can one make of the current discussions of the crisis in mathematical education (K-16) and how to deal with it? Everyone has his own view of mathematics, what it is, what it should be. Positions are advanced with a force and a conviction that has produced a scene reminiscent of the last act of Hamlet: most everyone is on the stage floor, dead, slaughtered by most everyone else. And hundreds upon hundreds of bloodstained letters to the editor, articles, texts, slides, videos, CD-Roms, teachers' guides, polemics, lie scattered about, both adored and ignored. What complicates matters enormously in the United States is the question of how to deal with classes of mixed backgrounds, a range of talents, and unqualified teachers.

I'll focus on just one type of discussion: the goals of mathematical education. The best analysis of the variety of possible goals, within my reading experience, is to be found in Paul Ernest's *The Philosophy of Mathematics Education*, 1991. Ernest distinguishes five types of mathematical educators, dubbing them the "industrial trainer," the "technological pragmatist," the "old humanist," the "progressive educator," and the "public educator." He then links up the goals of these various educators with various current social philosophies and philosophies of mathematics. To be sure, Ernest's background is education in the United Kingdom, and for the most part he is talking about secondary education, but a good fraction of what he writes is transferable to the US and to undergraduate college work.

Let me list some of the possible college goals:

- To create more research mathematicians

- To create more research scientists

- To create more technicians

- To act as a gate for admission to certain professions or simply to other courses

- To act as a course-of-study stuffer

- To teach enough to get get through day-to-day encounters with mathematics—principally money.

• To create a mathematically literate population that will be able to understand that the mathematizations that are put in place in society do not come down from the heavens, but are the work of people. To understand that these mathematizations are human arrangements and should be subject to the same sort of critical evaluation as are all human arrangements.

Focusing narrowly on mathematical content, the following subgoals can be distinguished as soon as one arrives at a knowledge of mathematics that is sufficiently deep: say, elementary calculus.

To instill an understanding and an appreciation for the:

major substantive and theorematic component

logical and deductive component

structural component

symbolic or formal component

numerical component

spatial component

pattern component

stochastic component

intuitive component

experiential component

algorithmic component

computational-symbolic component

applied component

experimental and inductive component

inventive component

discovery component

problem solving component

puzzle component

fun component

aesthetic component

natural language component

collaborative or group component

historical component

a sense of what creates value in mathematics

philosophical component

gender component

ethnic component.

All the above subgoals are valid. And who knows but that in the future other goals may develop. Is there a *via media*? The specific goals emphasized by a teacher or by the teaching materials used will create a certain type of student reaction and a certain "product." One supposes that a good mathematics teacher is aware of all of these components and manages implicitly to point out their existence to the class. It is clear also, that equal emphasis would be both impossible and undesirable.

The US is so large and has so many schools and colleges of different types, I see no great danger and I see a number of virtues if a wide variety of mathematics courses are taught. There is no need for the lockstep approach, as in some centralized educational systems, wherein one would know that on a given date, integration by parts is being taught in elementary calculus all over America.

Is there a danger that if this route is pursued that mathematics itself, touted as the universal language, will become Babelized? At the research level, it is already Babelized. We long ago reached the stage where two members of the same department, occupying adjacent offices, find little of a professional nature to discuss with one another. At the level of practical computer usage, it is Babelized. Is there, should there be, can there be one universal inter-comprehensible computer language?

Mathematics is an activity that is created by and supported by people with a wide variety of interests. No one group has or should have an iron grip on its creation, its dissemination, or its use.

Part XII

Echoes of Rousseau

Making Plans

As my discussions with Bernhelm and as my work on *Updating Jefferson* proceeded, I began to feel torn. I felt pulled in one direction by my love of mathematics and my concern for my profession, and I felt pulled in another direction by the realization that mathematics often promulgated a spirit and a view of the world that I found unacceptable.

By way of a historic parallel, I found myself appreciating more and more the conflicted feelings of the early evolutionists, Darwin included, many of them clergymen with an interest in geology or biology, asked themselves whether they would be able to abandon their time- and faith-hallowed belief in favor of what seemed to be scientific evidence. I've read that serious neuroses began to afflict some of these sensitive Victorian gentlemen. And I've noted on earlier pages how the German author Gerhardt Hauptmann (1862–1946) built this conflict into a tragic play.

I began to feel hesitant about going forward with the writing project. In the middle of my hesitancy, an event occurred that gave me a bit of a break. Over the centuries, the French have run many prize essay contests. In 1993, the French Academy once again announced a prize (60,000 francs) for the best essay on the topic: Have the sciences and arts purified morals? Bernhelm heard about it and told me. I encouraged him to make a submission, and he said "Well, let's do it jointly."

The First Contest. Fire: What is It?

In 1738 the French *Académie des Sciences* announced a prize for the best essay on the nature of fire. Voltaire entered. Later, Emilie du Chatelet entered the contest with an independent submission. Voltaire did not win. Emilie did not win. The prize was won by three people, one of whom was the Swiss mathematician Leonhard Euler (1707–1783).

Euler, to my mind, was the Mozart of Mathematicians. Tremendously brilliant and prolific, there is hardly a page of pure or applied mathematics to which he did not make significant contributions. Euler lived about the time that science was passing from talented and devoted amateurs (such as Voltaire or Benjamin Franklin) to inspired professionals. The awards committee was perceptive. In Euler's prizewinning "Essay on Fire," he regarded heat as a vibration of molecules and developed a vibratory theory. His submission was written in Latin and contains not a single mathematical equation! (*Dissertatio De Igne, in qua ejus Natura et Proprietates explicantur,* 1738. This work will appear in a modern edition in the near future, in Euler: *Opera, III/10,* a vast publication undertaking of the Swiss scientific community.)

The Second Contest. Science, Technology and Morals

Some years later, in late July, 1749, Jean-Jacques Rousseau, then thirty-seven and a citizen of Geneva, was walking from Paris to Vincennes where his friend Denis Diderot was jailed for publishing an essay on blindness that had atheistic overtones. He learned that the Academy in Dijon was offering a prize for the best essay on the topic: Has the restoration of the sciences and arts tended to purify morals? Rousseau's mind caught fire, and by the time he reached Diderot, he was beside himself with excitement. He wrote his essay, he won the prize (1750), and it made his reputation.

"Our souls," Rousseau asserted, "have been corrupted in proportion to the advancement of our sciences and arts toward perfection."

He based this uncomfortable judgement on several considerations.

(1) He cited historical situations as reasons for this uncomfortable judgement. First, he examined historical cases and found that while Sparta was pure, Athens was corrupt. The Persians were both moral and unenlightened. He also cited numerous other cases. In fact, said Rousseau, one may take this inverse correlation as a Universal Law of History.

(2) In a much repeated and anthologized assertion, Rousseau said that the sciences and arts owe their birth to vices:

> astronomy from superstition; eloquence from ambition, hate, flattery; geometry from avarice, physics from vain curiosity; and all, even Moral Philosophy, from human pride.

In a word, who would want to spend his life in sterile speculations, if each of us, consulting only the duties of man and the needs of Nature had time for nothing but the Fatherland, the unfortunate, and his friends?

(3) Rousseau claimed that the sciences and arts produce luxury, and that luxury is a source of corruption for morals as well as for politics. He claimed that the conveniences and luxury of civilization weaken the military virtues needed for self-defense.

What was bugging Rousseau? Was he sincere? Or was he simply a tease? Was he expressing dour Calvinism or Geneva provincialism? As a Swiss, did he know where he fit into the European scene? Was he a German intellectual? No? French? No? Was his downbeat view of humans due to a terrible childhood? In any case, he seems later to have repudiated the views expressed in his prize-winning essay.

Perhaps a more interesting question would be: what was going on in Europe that in a few short years, from 1738 to 1749, that changed the discussion from an inquiry about nature to doubts as to the consequences of such inquiries.

The Third Contest. Science and Morals, Once Again

In 1993, two hundred and fifty years after Jefferson's birth, and almost that many years after Jean-Jacques Rousseau wrote his prize winning essay, the French ministry of culture announced a prize contest on the same topic. The contest was open to all citizens of the European Community. Here is how it was stated: (my translation from the French)

ROUSSEAU CONTEST

In 1750, The Academy of Dijon raised the question whether the new burst of science and arts had contributed to the improvement of morals. J.J.Rousseau won the prize and developed this theme in his later work.

The situation today in Europe is, in certain respects, reminiscent of that of eighteenth-century France. The organizers of the prize contest, in once again proposing this subject, believe it offers today's essayists considerable opportunity for the expression of their opinions. A short, provocative and unpublished essay could open a new line of inquiry concerning the difficult societal relationship between knowledge, ethics, and politics.

Bernhelm, as I've said, heard about this contest. I suggested that he make a submission. He made a counter-suggestion that we make a joint submission. I thought about it for a while. I vacillated. I realized that though our views were close, they were hardly coincident. Bernhelm was a principled "Green" in a way that I am not. Bernhelm and his wife rode bicycles. I have a car. I find canned peaches a delight, while canned food of any kind is a no-no for Bernhelm.

In the end, I was spared a decision about collaborative participation in this contest because of yards of red tape. One of the conditions of the contest was that the entrants be members of the European Community. Bernhelm's wife Sussi came forward and said she'd volunteer to ring up the French cultural attaché in Copenhagen and ask whether it would be sufficient if, in the case of a joint submission, only one of the authors was a member of the European Community.

The attaché had no immediate answer. He said he would have to refer the matter to Paris and that would take time. The inquiry then resulted in such a flow of messages between Copenhagen and Paris, Paris and Copenhagen, Copenhagen and Brussels, and the attaché reported back in such ambiguous terms, that Sussi abandoned the inquiry and advised her husband he'd better go it alone. This was around the time that the EEC in Brussels promulgated firm rules and regulations as to what was and what wasn't a cucumber! Bernhelm agreed and this decision suited me perfectly.

In the weeks that followed, Bernhelm conceived of his entry as a series of five letters to the Ghost of Rousseau, in which a neo-Rousseauvian point of view would be expounded. It did not win the prize. But some of his points were well taken and several, are interesting to consider.

Bernhelm begins by stating

> I think that morality should be based firmly on what Rousseau called the two most simple operations of the human soul: "the one that interests us deeply is our own preservation and welfare, the other inspires us with a natural aversion to seeing any other being, but especially any being like ourselves, suffer or perish."

He continues addressing Rousseau:

> I believe that there was a time—including yours—when science played, on balance, a positive part in meeting two reasonable

human wants: first, by supporting the wider availability of adequate supplies of food, clothes, warmth, etc., and although scientific research was often motivated by military requirements or the desires of parasitic rulers, it contributed to the relief of mass poverty. Second, there was felt a need for knowledge, for enlightenment, for escape from a world of devils, of witches, and of inscrutability.

In your [Rousseau's] day, science still served to create a more coherent and intelligible world view by explaining a confusing variety of phenomena by means of a few fundamental concepts and principles.

Then, turning to the present, Bernhelm says that in the developed countries there are science believers who regard scientists as almost divine authorities and science skeptics who regard science as a cleverly run business for which scientists are agents. Both are mistaken:

The conventional opinion about science and technology is positive: that they created our wealth and comfort of life; that they have, to be sure, created certain practical and ethical problems, but that the solution to these problems is to be sought in a further scientific and technological development.

Both the believers' and the skeptics' attitudes towards science are forms of an intoxication induced by drinking a pernicious cocktail of shamelessness and ignorance, served up by the hi-tech science community and peddled by their agents in the media. Perhaps the shamelessness begins with boasting researchers habitually promising more than they can deliver. This prompts ignorant journalists to hold out miraculous prospects such as "Unlimited fusion energy just around the corner" and "Rocket transport for European weekend holidays in Australia."

I think that science is given too much credit for our present standard of living as if there had been no liberation from poverty, disease, and hard labour without science. Second, science and technology have produced spectacular results in this century, but I cannot see that our fundamental understanding of the world has made significant progress during the past 50 years.

A serious problem is that of uncontrollability:

> Every corner of modern society is occupied by complex, uncontrollable, unpredictable artifacts of science. That is dangerous enough, but still worse is that these artifacts create forces of society that are as inscrutable to an individual as in the dark ages when our forefathers humbly laid their destiny in the hands of their gods and gurus.
>
> Our reaction to the inscrutability of our world is much the same: we hand responsibility over to the scientific experts. As in the case of our forefathers, our chances for salvation would improve substantially if we trusted ourselves more and the higher authorities less.
>
> The gigantic scale on which both large and small industries contrive to implement mathematics and natural sciences in their production, their massive use of electronic equipment and modeling, allow them to overstep previous safety margins and run many processes closer to tolerance limits. This increases the potential for disaster if some gadget fails and something unforeseen happens. To claim that we can predict or control the consequences of the many processes that we set going is at best a self-delusion; the plain fact is that we are much better at innovating than at estimating and controlling the consequences of the innovations.
>
> Even more alarming than the physical threats to ourselves and to the environment is the fact that advances in science and technology lead to new kinds of mental and moral degradation. One of them is the adoration of science. This finds expression e.g., in the fact that the International Mathematical Union has declared the year 2000 the "World Mathematical Year" under the sponsorship of UNESCO. Its "First aim: the great challenges of the twenty-first century," to be envisioned by a committee of "first-class mathematicians!"

Ultimately, Berhelm envisions a reconstituted attitude towards science, a scientific Reformation:

> We need pragmatic thinking and to learn to be attentive to the consequences of our habits. Good intentions are not enough. We must turn back to the wisdom of the old Roman commanders:

QUIDQUID AGIS PRUDENTER AGAS ET RESPICE FINEM.

In English: whatever you do, act prudently and consider the outcome. Only that kind of pragmatic thinking may help us out of your [Rousseau's] dilemma:

"Discontented with your present condition for reasons which threaten your unhappy posterity with still greater vexations, you will perhaps wish it were in your power to go back."

Pragmatism may help us slow down the speed of innovation and in specified fields actually turn back to specified situations of the past. Nothing less will do.

And so, Mr. Rousseau, my answer to the question of whether the restoration of the arts and sciences has had a purifying effect upon morals is the same as yours: it has not.

Despite the similarity of some of these points from Bernhelm's letters to Rousseau, they deserve consideration. I quote a remark of André Gide: "Say it again. No one is listening." Finally, the judges, in their wisdom, did not award a first prize. Presumably, they thought that all the submissions were second rate.

The second prize, *proxime accessit*, as they say in British educational circles, was won by Guérgui Danailov, from Bulgaria, and his article can be found in *Commentaire*, No. 79, 1995. His article strikes me as a middle of the road piece of neo-Stoicism written by a person who had navigated through the perilous and tragic waters of Bulgarian communism. Danailov writes in conclusion (my translation from the French): "We are again torn between the 'yes' and the 'no.' Man has no solution other than to bow his head and to understand that this 'yes' and this 'no' coexist within himself."

There are conflicting ideals and you can't go for them all. This is Sir Isaiah Berlin's conclusion. It is very often my own. The judges were correct: it is second rate. The middle lane is the sensible lane, but it is never the fast lane.

Wherein I Write to Mr. Jefferson

As a counter in some ways to Bernhelm's letters to Rousseau, I found that I had to write to Jefferson, reechoing the latter's optimism.

Dear Mr. Jefferson:

When Jean-Jacques Rousseau submitted his prize-winning article to the academy, you were a boy of seven. Growing up, you came to love mathematics and the various sciences and you once wrote: "Nature intended me for the tranquil pursuits of science, by rendering them my supreme delight."

In fact, with all your involvement in public affairs, you found the leisure to practice architecture and to devise a number of practical mechanisms. In 1787 you wrote *Notes on the State of Virginia,* which detailed the natural and human resources of Virginia which then extended westward into *terra incognita.*

As President, you actively promoted the governmental support of science. You wrote eloquently about the positive role that science would play in fostering liberty, democracy, and the general well being of humanity in a new country that had replaced an aristocracy of birth with one of intelligence. Your thoughts ranged over the relationship between science and civilization at an elevated level.

Today, we are all children of the Enlightenment that you espoused. As consequences of these values, superstition, mysticism, irrationalism, dogmatisms, have abated. Tolerance, freedom of thought and of expression, education, information, have increased, though we see some possible limits looming. In the past two centuries, technology and science have moved forward at an accelerating rate; the life of the average person has been extended and made more comfortable. The face of the land and the contents of our minds have both been transformed in the process. Your thoughts and writings have served as guideposts for the future of our country.

Now where does mathematics fit into this?

Just as you, Mr. Jefferson, sent forth an expedition into the uncharted Louisiana Territory to take stock of what was there, today's scientists, technologists and sociometricians send forth their expeditions into space, into cells, into mathematical theorems, into physical theories, into statistical inquiries. They not only search for what is there but even create what is "there." Their findings and their creations are absolutely remarkable and would have been totally unpredictable to men like Laplace, Lavoisier, Humboldt or Gauss.

Since 1875, with the discovery and utilization of electromagnetic waves, and the development of atomic physics and of air and space travel, mathematics has become the standard way of describing what is. It is not only a standard means of determining what will be, but a frequently successful guide to the further exploration into what is potential.

Although, as you know, the prescriptive aspect of mathematics in a social context goes back at least to the US Constitution (controlling fractions in political bodies, the decennial census), social mathematizations now flourish beyond all expectations. Our daily lives, from buying groceries in the supermarket to responding to opinion polls, from automated buying and selling of stock to weather prediction, all these and more have been increasingly mathematized.

In your day, philosophers Hobbes, Locke, Rousseau—you read them all—put forward an idea, known as the "social contract," to explain the origin of government. Their "social contract" was conceived as a form of government organization. Prior to the contract there was, supposedly, a "state of nature" that was far from ideal. The object of the contract, as Rousseau put it, was "to find a form of association which will defend the person and goods of each member with the collective force of all, and under which each individual, while counting himself with the others, obeys no one but himself and remains as free as before." In this way, society improved lives that had been, as Hobbes put it, "solitary, poor, nasty, brutish, and short."

In our day, mathematics has proposed a variety of schemes—drive uniformly on the right side of the road, adhere to an agreed upon tax scheme, join an insurance pool, give one person one vote, cast by mail, or shortly from a computer terminal, compute and make policy based on the cost-of-living index. The agreement to adhere to such templates of behavior constitutes a social contract of a second kind. The new social contract, under which we now live, provides society and its members with a high degree of security but at the cost of some degrees of freedom.

My friend Bernhelm, in his letters to Jean-Jacques, emphasized the downside. Naturally, after the passage of so much time, Berhelm's reasons are rather different from Rousseau's. It would have been impossible for the philosophers of the Enlightenment to predict the consequences when such a large part of life has been "rationalized" via mathematics.

I believe that Bernhelm's message has been delivered to the world constantly since the end of World War II. The response to it has been up and down. The American response in 1974 to the gas "crisis" was positive and swift. Now new monster cars and trucks once again populate the streets. Ecological recidivism is rife, the concerns expressed by Bernhelm are still largely ignored by the majority, caught as it is by habits of thought, custom, law, and commerce. These bind it to archaic notions of what constitutes work, wages, wealth, health, or what even constitutes "life, liberty, and the pursuit of happiness." The notion that technology is the gateway to Paradise holds sway.

One looks around today, and one sees struggles between nations and smaller groups fought with the primitive passions of Scipio Africanus and Hannibal, with the pride and hot temper of characters described by Sir Walter Scott in his historical romances. They are fought not with the swords, spears, battle axes and war elephants of antiquity, but with the fearsome weapons of mass destruction provided by physics and technology. These weapons have been designed and executed with a tremendous assist from our mathematized civilization. When confronted by news of these struggles, I often feel like crying out: "Enough! Don't you know that science and technology have created a new material and psychological world. Discard your barbarous primitivisms. Come into the twenty-first century!" Alas, as Caliban said, "You have given me language and I have learned to curse."

The persistence of irrational violence has been a source of deep depression for those for whom scientific positivism had become the total basis for life. A positivistic, technological Eden has not yet emerged, and shows no signs of emerging. On the contrary, popular imagery in our contemporary literature and in our movies often depicts the immediate future as social primitivism flourishing within a high-tech universe. The scientists, technocrats, and social planners aided by information superhighways laden with statistics and by mathematized pollsters plugging into the heartbeat of the populace, would dearly like to say that science can rid us of these atavistic tendencies, and yet it would seem that science has been impotent to do so.

Reason as embodied in the Enlightenment led to (or did not prevent) the guillotine of the Reign of Terror. In the late nineteenth century, Germany led the world in theoretical science, but this did not prevent the enormities of Nazism. Marxism, said by its founders to be scientific—did not Marx even make some trenchant comments about the basics of the calculus?—led to or did not prevent the enormities of Stalin. The world scientific community of the period 1930 to 1950 led to and was not able to prevent Hiroshima.

The arrow of time, so thermodynamicists have assured us, is irreversible. Even those who believe that "acorns and roots" create a moral Eden cannot really go back to the past. Even as they reject cars for bicycles, bicycle travel has been made possible by the paved roads designed for cars. We cannot forgo the advances we have made in medicine, even if we wonder what the possible consequences would be if humans were as long lived as sequoia trees. Whether a can of peaches is energy-efficient or whether it robs us of a dimension of life by denying us the delights of our own trees in their season, we are unlikely to give up what we have learned about efficient and sanitary food production and distribution. We are addicted to our computers even when we wonder whether they will replace literacy as we have understood it by a new kind of communication and a new kind of education.

There are those, Mr. Jefferson, who say that while it is true we cannot go back, we must go forward by restoring some aspects of the past. The advocates of this point of view cite a number of successes. The rapid increase just mentioned in the efficiency of automobiles at the time of the oil crunch was a remarkable achievement. The gradual restoration of fish to some of our waters and the wild turkey to our New England woods is encouraging. The general increased awareness—even if painfully slow—of the finiteness of the planet and of our tremendous potential for self-destruction is heartening.

The possibility of retrogression is real, as you yourself were aware. On June 18th, 1799, you wrote a long letter in answer to a young man who had inquired of you what mathematics and what science he ought to know to be considered educated. After detailing specific subjects, you went on to praise science in general, to speculate both on its endless potentiality and on a possible turning away from it. You wrote to him:

> I join you therefore in branding as cowardly the idea that the human mind is incapable of further advances. This is precisely the doctrine which the present despots of the earth are inculcating & their friends here re-echoing; & applying especially to

religion and politics; & that it is not probable that anything better will be discovered which was not known to our fathers. We are to look backwards then & not forwards for the improvement of science, & find it amidst feudal barbarisms and the fires of Spital-fields, but thank heaven the American mind is already too much opened to listen to these impostures.

The generation which is going off the stage have deserved well of mankind for the struggles which it has made, & for having arrested that course of despotism which had overwhelmed the world for thousands and thousands of years. If there seems to be danger that the ground they have gained will be lost again, that danger comes from the generation your contemporary. But that the enthusiasm which characterizes youth should lift its parricide hands against freedom and science, would be such a monstrous phaenomenon as I cannot place among possibilities in this age & this country.

If you were giving advice today to a young person asking about education, I think you might very well have included a thought that was admirably expressed by Mogens Niss, professor of mathematical education in Denmark:

> It is of democratic importance to the individual as well as to society at large, that every citizen is provided with instruments for understanding the role of mathematics in society. Anyone not in possession of such instruments becomes a "victim" of social processes in which mathematics is a component. The purposes of mathematics education should be to enable students to realize, understand, judge, utilize, and perform the application of mathematics...in situations which are of significance in their private, social and professional lives.

Freedom and science. How strongly linked in your mind they seem to be. Rather less so in ours. It would make a most interesting project, Mr. Jefferson, to comment on your long letter to the young student sentence by sentence in the light of the two centuries that separate us from it. But I shall limit myself to one point.

For you, science was an embodiment of morality. You would not agree with the assertion often made that science and technology are morally neutral: "the gun does not shoot by itself." To put it differently: science and technology do not exist as ideal objects in the world. They are created, developed and applied by people. As such, their creation and their pursuit, their support and encouragement by society express a certain human intent. What one wants is critical awareness of this intent and anticipation of how this intent may be diverted toward the immoral by the inner nature of the technological objects created.

One might object that the instances of optimism that I cited a moment ago (such as the revitalization of our rivers) are mere band-aids applied to a natural world that is hemorrhaging badly. Perhaps. I do not have such a low estimate of these attempts because experience has shown that draconian or truly revolutionary solutions can also be a source of great difficulties. To lift our "parricide hands against freedom and science" would be, I agree, a disastrous event.

There is yet, Mr. President, another reason for my optimism, different from yours. I was reminded of it recently as I was browsing in the stacks of my university library. I chanced to come across *Dangerous Thoughts*, a collection of essays written in 1939 by Lancelot Hobgen, a biomathematician and a very successful popularizer of mathematics and its applications.

Published on the eve of World War II, its opening sentence reads: "My Marxist friends assure me there are now only two choices: Communism or Fascism." History shows that civilization can lock itself into believing and accepting dichotomies; either this or that: *aut Caesar aut nihil*. The world is either scientific or humanistic, capitalist or communist; people are either intelligent or stupid; sane or insane. Mighty thinkers have gone along with such splits; have created them. The tragic potential of either/or is great. In specific cases, it may take centuries to loosen the grip of the either/or. Mathematical thinking with its razor-sharp true or false, 0 or 1 is particularly prone to this kind of thinking.

Can we split life neatly into two categories: the first asserting Enlightenment values while the second view asserts their denial? Now since you are living in the Elysian fields with knowledge revealed, tell me what guarantees that this split exhausts all the possibilities?

On the contrary, it seems to me that the history of ideas suggests that the intellectual dichotomies of the past weaken and become irrelevant as they dissolve and reform into other issues. The dogmatic synthesis of the thirteenth century gradually gave way under the force of absolutely new events: the opening up of the New World, the opening up of larger physical and biological worlds via the perfection of powerful instruments of observation, the desire to look and to believe what one saw.

Even as one looks, what one sees is that *panta rhei*, everything flows; in time, any specific either/or is destroyed. The present is poised ambiguously. There is a faith that denies the either/or and that asserts the possibility of braving the flow. Bolstered by that faith, we live "on the windy side of hope." As my friend Clara Claiborne Park has written "Hope is...to be practiced whether or not we find it easy or natural, because it is necessary to our survival as human beings."

Yours sincerely,

PJD

Part XIII

Still Carrying On

On the Strudelhofgasse:
Standing Room Only

June, 1997 was beautiful and I was back in Vienna for a short visit. Off the plane and into a light outdoor supper with Liese and Karl, both mathematicians, in the amusement section of the Prater.

"What would you like to do on this visit?"

"Nothing special. Talk to people. Relax. Talk to Franceska."

"Why Franceska?" Liese asked me.

"Well, the last time I was here, she told me that she had been talking to Pythagoras, and I think that might be the basis of a story for me."

"She's weird, I have to tell you."

"I know. You don't have to tell me."

Karl told me that on Wednesday there would be a public lecture by the renowned Oxford polymath Sir Roger Penrose, and did I want to go?

"Of course. Absolutely. Incidentally, what's Penrose talking about?"

"The mind. The body. Consciousness. Gödel's theorem. Computers. The Second Law of Thermodynamics. Physiology. Quantum physics. The limits of thinking. Is the mind a computer? But it's to be a popular lecture."

Karl met me at my hotel and we took the tram to the University of Vienna and arrived about a half hour before Sir Roger's talk. The hall was packed, and we just managed to find seats in the third row. Penrose's talk had been widely advertised in the newspapers, and the whole of Vienna, so it would seem, wanted to see and listen to this scientific lion. Students in their grubby clothes (a worldwide phenomenon?); members of the academy; faculty, some wearing suits, some in T-shirts; fashionable ladies; townspeople.

Soon the aisles were filled and soon after that people sat on window sills, jammed the interstices and overflowed into the outside corridor. I saved a seat next to me for Liese, and she showed up before I was embarrassed. Two seats over, I saw Professor Jack Todd, my old boss at the National Bureau of Standards, and teacher at the California Institute of Technology of many mathematical and computer luminaries.

National TV cameras, videos, recording equipment, and overhead projectors were all in place (Penrose was a master at handling classroom media).

I looked around and high up on the left hand wall I spotted a memorial tablet. It said "*Hier Wirkte Kurt Gödel Von 1932–1938*" (Gödel worked here from 1932 to 1938). I took a photo of it. Another tablet memorialized other famous mathematicians and scientists who had been at the University of Vienna: Alfred Tauber, Hans Hahn, Johann Radon, Erwin Schroedinger. More ghosts heard from.

Penrose spoke. Mathematics, thinking, consciousness. Three worlds: platonic mathematics, the world of ideas, the physical world. He displayed circles to symbolize the worlds. There is a fourth world: the world of people and their interactive relationships. (I thought to myself: both Gödel and Franceska had another world beside these: the world of mystical illumination.) Penrose displayed physical diagrams, showed the non-periodic tilings he had discovered. Said that details would be given in tomorrow's (technical) lecture. In the midst of the talk, a thunderclap outside. The speaker paused, laughed. The audience laughed. Was the physical world agreeing with what the speaker had just said? Was Jupiter Pluvius commenting on Penrose's Worlds?

I didn't understand Penrose's talk. Other people I spoke to also admitted their difficulties. The speaker possessed virtuoso knowledge across many fields that I did not, and had a flair for making bold connections and conjectures that I could not follow. Perhaps he's on to something, I said to myself, but do we understand the basics of quantum mechanics? Do we understand the basics of consciousness?

Ultimately the talk ended, and after a few questions from the audience had been answered, (Sir: is beauty an element in a theoretical construction?) we filed out of the hall slowly. I began to hear very loud sounds of the latest popular music blasting away from distant boom boxes and drowning out whatever echoes of the lecture might still have been reverberating. Students, in from the rain, were milling about and buying food and soft drinks from a stand set up near the door of Penrose's lecture room. "Traditional end of the term celebration," Karl explained. "The students were cautioned not to start their music until the lecture was over."

The new generation, I thought, for better or for worse, ingests mathematics unconsciously in its recreations: digital recordings, data compression based on the Fourier Transform and wavelets, programs of numerical analysis. In my bag, by way of contrast to the output of the boom-box, was a disk of Schubert's *Lieder* sung by Fischer-Dieskau, given to me as a gift by Dr. Margarete Mahrenburg, a historian, who works at the Austrian Academy. "This," she said to me, "is the true, the prime Vienna," and I wondered for how long and in what circles or worlds that appraisal would remain valid.

Small talk overheard in the corridor:

"Did you know that so-and-so (a philosopher of science now in the States) hides his Viennese origin?"

"Why does he do it?"

"Who knows? An unsolved problem."

Liese, Karl, and I took the tram to *Zur Schnellen Schildkroete*, a popular restaurant not too far from the university. There we were joined by their friends the Mahrenburgs, who had also been at Penrose's presentation. The table conversation turned to Franceska and her conversations with Pythagoras.

"Did you know," Margarete Mahrenburg asked me, "that Hermann von Keyserling played around with Pythagoreanism? No? You didn't? Then I'll send you a reprint."

"Do that," I said. I didn't know who H. v. K. was but I was willing to learn.

I ordered *Rindsuppe* and a *Wienerschnitzel*. I was delighted to see that the schnitzel came with the canonical sardine on top. I had forgotten about the sardine. Afterwards, I had a mélange, a popular coffee mixture. Life is, always has been, an ambiguous mélange. Make it precise, if you can; disambiguate it before tucking in at night. But for what purpose?

"What are you working on these days?" I asked Margarete.

"Voltaire, the scientist."

"I didn't know Voltaire was a scientist!"

"Oh yes, he and Emilie conducted regular experiments in a laboratory they had set up so as to verify Newton's Laws."

"Emilie? Who's she?"

What I found out I've already told in the chapter on the French Enlightenment.

How I Used Blackmail To
Become a Professional

The International Congress of Mathematicians meets every four years in different parts of the world. In August, 1998, it was the turn of Berlin, Germany, to host the Congress. Berlin, reunited, rebuilding rapidly, (six thousand building cranes throughout the city), with the government offices returning from Bonn, was seething with pride, excitement, boosterism, and problems. More than three thousand mathematicians from all over the world were expected. Berlin wanted to put on a great show and it did.

I was asked by Professor Gottfried Mueller, of one of the Berlin universities and a member of the arrangements committee, to give a Urania Talk in conjunction with the congress. Of course I was flattered by the invitation, and when I wrote back saying that I would be glad to give a Urania talk, I added: please explain to me just what a Urania talk is or should be. Tell me also who the audience would be.

Mueller answered me that a Urania talk was a popular talk and that the audience would be interested Berliners together with Congress members who happened to be attracted by my subject. I thought Mueller's answer was a bit vague, but I let it go at that, not, however, before asking him for any suggestion he might have on what would constitute an appropriate topic. I might have anticipated Mueller's answer. It was the standard answer a speaker gets whenever he asks his host what he should talk about: the host is sure that whatever the speaker wants to talk about will be just fine. I'm not so sure about that.

My first idea was to title my talk "Is Thinking Obsolete?" the concept being that in a multimedia age, an age in which we are absolutely drowning in information, the tendency (already noted in some quarters) to replace thought by information retrieval would increase disastrously. The title was surely an eye-opener or an eyebrow raiser, but on consideration, it struck me as too brutal and too general for a mathematically inclined audience. So I altered the title to "The Prospects for Mathematics in a Multimedia Civilization," a bit long, a bit cumbersome, but at the time, I couldn't come up with a punchier title.

Mueller had arranged by e-mail that right after the opening session of the congress in the International Convention Center of Berlin we should meet at such and such a place so that he might brief me on some of the details of the Urania set-up. When I managed to get down from the far right field bleachers (so to speak) of the vast convention center, our designated meeting place (*Treffpunkt*, in German) was so crowded with thousands of wandering mathematicians eating a complimentary lunch of Berliner meatballs and other savories, that I couldn't find Mueller nor he me. Added to the difficulty was the fact that I had no idea what Mueller looked like, nor he me.

The opening session took place on a Tuesday and I was scheduled to speak on the following Friday night. Perhaps I didn't really want to meet Mueller lest he give me more precise Urania information that would force me to recast my whole speech.

A small reception at the Urania was planned for Thursday night. Mueller would be there to say a few words. I could meet him there and he could clue me in for Friday. I would have a day in which to adapt my talk for the audience.

The Urania is a theater located in central Berlin within walking distance of my hotel, just past the famous *Ka De We* department store and a few blocks down the street east of the *Wittenbergplatz* underground station. (Incidentally, Urania, in German, is pronounced Oorahniya, and has nothing to do with fissionable materials. Urania was the muse of astronomy.)

And what did I find when I got there on Thursday? The theater was very large, much like any theater on Broadway, with a marquee, a lobby, and box office. But inside there was also a gift shop, a mezzanine area for pastry, coffee and wine, a museum area, lecture rooms, a concert hall, and facilities for adult education and arts and crafts.

The reception was held in the mezzanine. A small band was playing—much too loudly for my ears trained to Fred Waring. I sat down at a table. I was soon joined by three younger friends from Berlin and Vienna. We ate some kind of sweet cheese thing. I drank coffee; they, wine.

My friends told me—I did not know this in advance—that at all Urania talks admission is charged. I asked whether the members of the International Congress would get in free by showing their badges. They answered no. Everyone had to pay six Deutschmarks. (About $3.50; not a tremendous sum.) They also hinted it would be cool if I, as speaker, could get them complimentary tickets for my talk tomorrow. (Q.: *Wie sagt man "cool" auf Deutsch?* Ans: "*Cool.*") I allowed as how it would be cool, and I promised them some tickets. Do you know Mueller, I asked my friends, he wants desperately to see me. They did not know him.

The band ceased playing. A lectern was set up and words were spoken. I sensed that the last speaker was my man Mueller. When he finished, I corkscrewed my way through the crowd up to the lectern.

"Hello. I'm Davis. I'm sorry I missed you the other day at our *Treffpunkt.*"

"Yes. I'm sorry, but I was there. I'm very glad to know you."

We shook hands.

"What I wanted to tell you. It's of not great importance," Mueller continued, "Only this. Tomorrow night, come early. Go over to the gift shop and make yourself known. They will take you to the office of Dr. Schmidt, the director of the Urania. Dr. Schmidt will show you what you have to know."

We talked on for a bit, touching on the usual polite and civilized nothings. Then I took courage.

"I understand," I told Mueller, "that admission will be charged for my talk."

"That is correct."

"You know, this is the very first time in my life that admission has been charged to hear me speak."

"Well, I suppose there is a first time for everything," Mueller said, managing a smile.

"And I suppose that in that case, I might call myself a professional speaker."

Mueller still smiled and I continued.

"I've heard that professionals always get a few complimentary tickets. At least this is what I read in biographies. Three close friends are coming to my lecture and I would like complimentary tickets for them. Can you get them for me?"

Mueller thought a moment, and replied rather severely, "I'm afraid there is no way. Everyone has to pay for a Urania lecture."

When I heard the words "no way," I'm not sure what devil got into me, or what act of foolish bravado overcame me, but a second soul that I never realized resided in my breast took charge, and I said to Mueller in no uncertain terms, "In that case, I will not speak tomorrow evening."

Mueller turned red, yellow, green, all the colors of a stop light. Perspiration broke out on his forehead. He saw his careful planning of months going down the drain. He visualized his embarrassment in arranging for a speaker who was in Berlin but who would not speak.

Perspiration metamorphosed into inspiration.

"Just wait a moment," Mueller said, "don't move. Just stand right here."

I waited in the throbbing crowd and high-decibel band music. In five or so minutes, Mueller was back. He handed me three tickets.

"Here are your tickets. Complimentary."

I smiled and we parted on the best of terms.

Ten minutes later I found my friends and gave them the tickets.

"I'm now a real professional," I told them, "I have complimentary tickets to distribute."

I did not yet realize the full extent to which I had gone pro. That realization would come on the day of my lecture.

On Friday I planned to walk from my hotel to my lecture so as to arrive there 45 minutes early. A light rain was falling, so I hustled even more. Arriving in the vicinity of the Urania, from across the street I saw my name up in lights on the marquee: "Phil Davis/ The Prospects of Mathematics."

Up in lights! In one of the great cities of the world. My head whirled. Up in lights with Lawrence Olivier, Richard Burton, Ralph Richardson? Phil Davis in "Hamlet." Why not? Why limit myself to mathematics? A new career at the age of seventy-some-odd? But no, I thought, not in "*Hamlet*" but in "*Cyrano de Bergerac*." Why not, indeed? That role would be quite appropriate. (I would need a stand-in for the duel scenes.)

> To sing, to laugh, to dream
> To walk in my own way and be alone
> Free with an eye to see things as they are
> To cock my hat where I choose...

The illusion of treading the boards lasted a few seconds. I crossed the rainy street, presented myself at the gift shop, and was led to the office of Dr. Schmidt, the director of the Urania.

"Ah, Prof. Davis. *Sehr angenehm*. Come in. Come in and relax. You have lots of time. We can talk a bit."

I entered Dr. Schmidt's ample, comfortably furnished office and sat down. His walls were covered with photographs of many men and women. I assumed they were some of the great names who had performed at the Urania. Schmidt verified this. I recognized Thomas Mann and Niels Bohr. Wow! The arrangement was reminiscent of Lindy's restaurant in the theater district of Manhattan, whose walls were covered with the signed photos of all the stars of the world of entertainment.

Dr. Schmidt told me the history of the Urania. It was established more than a century ago to provide popular lectures in science, literature, and the arts. It also serves as an adult education center, and from time to time as a special exhibit museum. The International Congress of Mathematicians, in arranging for a number of lectures ancillary to the Congress itself, rented the theater for a flat fee. The six-mark admission charge went to the congress and not to the Urania. All now stood revealed.

Lecture time arrived. I spoke in English, of course, and told my audience right at the beginning that I would speak very slowly. The lecture went well. I had a prepared text, but I ad-libbed as I usually do. It would be a tricky business to make spur-of-the moment jokes with a transnational, translinguistic, diachronic audience—but they worked.

Pros eat after their performance, not before. I was ready now to do the same. Together with the three recipients of the complementary tickets and a couple of others, I walked to an Italian restaurant in the nearby *Wittenbergplatz* where I loaded up on Spaghetti Bolognese, a very popular German dish.

How did Prof. Mueller wangle the complementary tickets? I assume he simply bought them out of his own pocket. It bothers me and I intend to reimburse him—when I get around to it.

What follows now is my Urania Talk, fleshed out rather a bit, but with all my ad-libs subtracted.

The Prospects for Mathematics in a
Multi-media Civilization

Poincaré's Predictions

Ninety years ago, at the Fourth International Congress of Mathematicians held at Rome in 1908, Henri Poincaré gave a talk entitled *"The Future of Mathematics."* He mentioned ten general areas of research and some specific problems within them, which he hoped the future would resolve. What strikes me now in reading his article is not the degree to which these areas have been developed—they have—but the necessary omission of a multiplicity of areas that we now take for granted and that were then only in utero, or not even conceived.

Though the historian can always find in the past the seeds of the present, particularly in the thoughts of a mathematician as great as Poincaré, I might mention as omissions from Poincaré's vision, the intensification of the abstracting, generalizing and structural tendencies, the developments in logic and set theory, the pattern-theoretic, the emerging of new mathematics attendant upon the physics of fluids, materials, relativity, quantum theory, communication theory. And of course, the computer, in both its practical and theoretical aspects; the computer, which has altered our lives almost as much as the "infernal" combustion engine, and which may ultimately surpass it in influence.

Poincaré's omission of all problems relating immediately to the exterior world—with the sole exception (!) of Hill's theory of lunar motion—is also striking.

How then can the predictor with a clouded vision and limited experience proceed? Usually by extrapolating current tendencies forward linearly.

What Will Pull Mathematics into the Future?

Mathematics grows from external pressures and from pressures internal to itself. I think the balance will definitely shift away from the internal, and there will be an increased emphasis on applications. Mathematicians require support; why should society support their activity? For the sake of pure art or knowledge? Alas, we are not classic Greeks or eighteenth-century aristocrats, and even their material was pulled along by astronomy and astrology and geography. Society will now support mathematics generously only if it promises bottom-line benefits.

Now focus on the word "benefits." What is a benefit? In a famous epigraph to his book on scientific computation, the late Richard Hamming of the old Bell Telephone Laboratories said "The object of computation is not numbers but insight." Insight into a variety of physical and social processes, of course.

But I perceive (forty years later and with a somewhat cynical eye) that the real object of computation is now neither numbers nor insight, but to make money—often via computations authorized by project managers who have little technical knowledge. If, by chance, humanity benefits, then so much the better; everybody is happy. And if humanity suffers, then some people will cry out and form chat groups on the Web, or hackers will attack computer systems or humans; while techno-utopians explain that you can't make omelets without breaking a few eggs.

Pure mathematics will move closer to applications, while justifying its purity to the administrators, politicians and the public with considerable truth, saying that one never knows in advance what products of pure imagination can be turned to society's benefit. Employing that most weasel of rhetorical expressions: "in principle"; in principle all mathematics is potentially useful. I cite a few examples.

The physical and engineering sciences begot and nourished the development of what was probably the major advance in mathematics since the 1600s: calculus and its offshoot differential equations, or as one might say simply, continuous mathematics.

Theories of electrodynamics, of fluids and plasmas, of elastic and plastic solids, of dynamical systems, atmospheric and oceanic modeling are written in the language of the calculus.

Fractals appear in geophysics and geomorphology. Stochastic (i.e., probabalistic) differential equations find their way into economic theories of securities.

Graph theory, topology, probability are used in image analysis and mathematical morphology.

Boolean matrices, combinatorics, difference equations are vital to ecology and sociometry. Time series analysis appears in automatic medical diagnostics. Algebraic group theory has been in quantum physics since the Twenties. Logic and a variety of algebraic structures are employed in theoretical computer science. The theory of numbers is indispensable to cryptography.

I could go on and on, detailing the specific branches of mathematics that are used in the life sciences, in military operations, in law, in politics, in film animation and computer art, in vending technology such as product striping, but I think I have made the point.

All of these applications converge in mathematical education and its methods. A colleague has written me:

> My teaching has already changed a great deal. Assignments, etc., go on the web page. Students use e-mail to ask questions, which I then bring up in class. They find information for their papers out there on the web. We spend one day a week doing pretty serious computing, producing wonderful graphics, setting up the mathematical part of it and dumping the whole mess into documents that can be placed on a webpage. I am having more fun than I used to, and the students appear to be having a pretty good time while learning a lot. Can all this be bad?

The classic modes of elementary and advanced teaching have been amplified and sometimes displaced by computer products. A good computer store has more of these products for sale than there are brands of cheese in the famous *Ka Da We* department store in Berlin. Will flesh-and-blood teachers become obsolete?

Is all this perceived as good? Apparently not always. Another colleague has written me:

> On balance, I believe that science will suffer in the multi-media age. My experience is that true thinking now goes against the grain. I feel I have to be rude to arrange for a few peaceful hours a day for real work. Saying no to too many invitations, writing short answers to too many e-mail questions about research. A letter comes to me from a far corner of the world: "Please explain

line six of your 1987 paper." I stay home in the mornings hiding from my office equipment.

Big science projects, interdisciplinary projects, big pushes, aided and abetted by multi media and easy transportation have diminished my available time for real thought. I am also human and succumb to the glamour of today's techno-glitz.

A collaborator working with me and much in love with mathematical databases, picked up (after clever filtering) more than 100,000 references to a key word that was relevant to our work. This produced an immediate blockage or atrophy of the spirit in him. We wondered whether we could afford the time to assess this raw, unassimilated information overload or should simply plow ahead on our own as best we could.

Semioticist and novelist Umberto Eco wrote (In *How to Travel with a Salmon*), "the whole information industry runs the risk of no longer communicating anything because they tell too much."

Nonetheless, my eyebrows were raised recently when I learned that as part of a large grant application to the National Science Foundation, the applicants were advised to include a detailed plan for the dissemination of their work. In the multi-media age, mathematics is being transformed into a product to be marketed like other products.

Thus, every aspect of our lives is increasingly being mathematized. We are dominated by and we are accommodating to mathematical machines and to the arrangements they prescribe. Yet paradoxically, the nature of technology makes it possible, through chipification, for the mathematics itself to disappear into the background and for public awareness and understanding of mathematics to diminish even as its presence hovers spectrally over society.

It is probably the case that despite the claims of educational administrators, the general population needs now to know less mathematics than at any time in the last several hundred years. What civilization needs is a critical education that brings an awareness and judgment of the mathematics that dominates it, and is able to react with some force to evaluate, accept, reject, slow down, redirect, reformulate the abstract symbols that are affecting their concrete lives. Technology is not neutral. It fosters certain kinds of behavior. Mathematics is not neutral.

Individual mathematicians are aware of this. Groups calling themselves "technorealists" have web sites. But with some exceptions, the awareness has not yet penetrated the educational process.

Benefits

All these developments will surely have benefits; and the advancing waves of new technologies that are sweeping over us hardly need a public relations agent to trumpet them. The size of some personal fortunes in the computer business derives from the public saying "Yes" to all this. "Bring on more." The mental gridlock has only begun to appear.

The Inner Texture (or Soul) of Mathematics

Let me now go up a metalevel and ask: how will multi-media affect the conceptualization, the imagery, the methodology of future mathematics? The metaphysics or philosophy of the subject? What is mathematics going to do for the world outside mathematics? This is for me a tremendously interesting but difficult question. Here again, all I can do is to describe what I see and to project forward. I shall make my attempt in terms of what I call "The Tensions of Texture."

The Discrete vs. the Continuous

William Everdell in the *The First Moderns* says that to be up to date is to be discrete, to be discontinuous. Everdell shows how this has operated in literature and art as well as in science and mathematics. Not so long ago, there was a movement afoot in the US, asserting that continuous mathematics ought to give way in education and in philosophy to discrete mathematics. This movement seems to have quieted down a bit.

The Deterministic vs. the Probabilistic

This split not only applies to the modeling of the exterior world, but resides interior to mathematics itself. There are now probabilistic demonstrations within the corpus of mathematics. Are the truths of mathematics therefore probabalistic? How can we live with such truths?

At the same time, it may be observed that one little old lady residing in Rhode Island, winning $23,000,000 in the powerball lottery, does more to question and destroy the relevance of mathematical probability theory in the public's mind than all the philosophical skeptics, myself included.

The two dichotomies mentioned: continuous/discrete, determinstic/probabilistic, are old but they persist. The question of determinism goes back surely as far as the philosophic discussions of "free will."

Then there are new dichotomies.

Thinking vs. Clicking

I have heard over and over again from observers that "Thinking increasingly goes against the grain." Is thinking becoming obsolete? To think is to click. To click is to think. Is this the equation for the future?

The mathematician/philosopher Alfred North Whitehead wrote in one of his books that it was a mistake to believe that one had constantly to think. Do not the rules, the paradigms, the recipes, the algorithms, the theorems and generalizations of mathematics reduce the necessity for thought? Did not Descartes write that his specific goal was to bring about this condition? And in the historic past was not thinking confined to a special class of people? Have not thoughts over the whole of history been controlled—otherwise one might be declared a dangerous heretic or a traitor? Were not women forbidden to think and to study?

Experimental mathematics, visual theorems, are increasing in frequency. There are now two types of researchers: the first try to think before they compute; others do the reverse. I cannot set relative values on these strategies.

Ernest Davis, a researcher in artificial intelligence, has written me: "Your question 'Is Thinking Obsolete?' is very much to the point. This has certainly been the trend in AI over the past ten years (just now beginning to reverse itself)—trying accomplish things through huge brute-force searches and statistical analyses rather than through high-level reasoning."

We can also ask in this context, is traditional advanced mathematics obsolete? For example, what portions of the theory of differential equations retain value when numerical solutions are available on demand; when, in many instances, computation is far more successful in explaining what is going on than is classical analytic or theorematic mathematics?

Words or Mathematical Symbols vs. Icons

A semanticist, Mihai Nadin, has written a large book, *The Civilization of Illiteracy*, on the contemporary decline of the printed word; how the word is being displaced by the hieroglyphic or iconic mode of communication. There is now computer-induced illiteracy and innumeracy.

There is no doubt in my mind that this displacement will have a profound affect on the inner texture of mathematics. Such a shift already happened four thousand years ago. Numbers are among the oldest achievements of civilization, predating, perhaps, writing. In his famous book *Vorgreichischer Mathematik*, Otto Neugebauer explains how hieroglyphs and cuneiform are written, and how this affects the forms of numbers and the operations with numbers. Another such shift occurred in the late middle ages when algebraic symbolisms began to invade older texts.

Mathematics as Objective Description vs. Mathematics by Fiat. Or: The Ideal vs. the Constructed and the Virtual

Applied mathematics deals with descriptions, predictions and prescriptions. We are now in a sellers' market for all three. Prescriptions will boom. There may indeed be limits to what can be achieved by mathematics and science (there are a number of books on this topic), but I see no limits, short of the willingness of humans to endure them, to the number of mathematizations that can be prescribed and to which humans are asked to conform.

In the current advanced state of the mathematization of society and human affairs, we prescribe the systems we want to put in; from the supermarket to the library to the income tax to stocks and bonds to machines in medical examination rooms. All products, all human activities are now wide open to prescriptive mathematizations.

Prof. David Mumford anticipates a great increase in the invention of new mathematical structures. The potentialities and the advantages envisaged and grasped by the corporate world will lead it to pick up some of the developmental tab. As it does, the human foot will be asked, like Cinderella's sisters' big feet, to fit the mathematical shoe. If the shoe does not fit: tough for the foot.

What Is Proved vs. What Is Observed

This is the philosophical argument between Descartes and Giambattista Vico. I venture that as regards the generality of users of mathematics, its proof aspect will diminish. Mathematics does not and never did belong exclusively to those who call themselves mathematicians and who pursue Mathematics with a capital M. I would hope that the notion of proof will be expanded so as to be acknowledged and presented as one part of the larger notion of mathematical evidence.

Euro or Western Mathematics vs. Other National or Ethnic Mathematics

The present corpus of mathematical experience and education has come under attack as a result of social-political forces. Today's ethno-mathematicians remind us that different cultures, primitive and advanced, have had different answers as to what mathematics is and how it should be pursued and valued. (e.g., ancient Asian mathematics was carried on in a proof-free manner. Ancient Indian mathematics expressed itself in verse. Other cultures have expressed in weaving patterns, things that Euromaths perceive as having mathematical interpretations.) In my view ethnomathematics is appropriate to anthropological or historical studies, but it is as irrelevant to the current study of mathematics as Gandhi's spinning wheel is to the economics of India.

Male vs. Female Mathematics

Mathematics has been perceived as an expression of male machismo. Margaret Wertheim is a TV writer as well as a former student of math and physics. Let me quote from her recent book *Pythagoras's Trousers*:

One of the reason more women do *not* go into physics is that they find the present culture of this science and its almost antihuman focus deeply alienating. After six years of studying physics and math at university, I realized that much as I loved the science itself, I could not continue to operate within such an intellectual environment. (p. 15)

Wertheim maintains that if more women were in mathematics and science (particularly in physics), they would create

an environment in which one could pursue the quest for mathematical relationships in the world around us, but within a more human ethos.... The issue is not that physics is done by men, but rather the *kind* of men who have tended to dominate it. Mathematical Man's problem is neither his math nor his maleness *per se*, but rather the pseudoreligious ideals and self-image with which he so easily becomes obsessed.

More women are entering mathematics and science, and it will take at least two generations to observe whether or not Wertheim's vision will materialize and what it implies.

The Apparent vs. the Occult

In a rather different and disturbing direction, we have the concern on the part of some mathematicians and physicists with hermeticisms, apocalypses of various sorts: final theories of everything, secret messages hidden in the Bible, everything under the sun implied by Gödel's theorem.

The old marriage of literacy and rationality, in place since the Western Enlightenment, seems to be ending in divorce. Western rationality has shacked up with fanaticisms. There is a long history of stoicism ending up with the occult.

Are these part of the breakdown of a literate civilization or merely the age-old and temporary anxiety that accompanies the arrival of a new millennium?

Soft Mathematics vs. Traditional Mathematics

I have picked up the term "soft mathematics" from Keith Devlin's popular book *Goodbye, Descartes*, which describes the difficulties of the relationship between natural language, logic, and rationality. These difficulties, Devlin asserts, cannot be overcome by traditional mathematics of the Cartesian variety, and he hopes for the development of a "soft mathematics" —not yet in existence—that "will involve a mixture of mathematical reasoning, and the less mathematically formal kinds of reasoning used in the social sciences."

Devlin adds that "perhaps most of today's mathematicians find it hard to accept the current work on soft mathematics as 'mathematics' at all."

Nonetheless, some see the development as inevitable, and Devlin uses as a credentialing authority the mathematician-philosopher Gian-Carlo Rota. Rota comes to a similar viewpoint through his phenomenological (Husserl, Heidegger) orientation.

After listing seven properties that phenomenologists believe are shared by mathematics (absolute truth, items (not objects), nonexistence, identity, placelessness, novelty, rigor), Rota goes on to say:

> Is it true that mathematics is at present the only existing discipline that meets these requirements? Is is not conceivable that some day, other new, altogether different theoretical sciences might come into being that will share the same properties while being distinct from mathematics? - "Ten Remarks on Husserl and Phenomenology," Address delivered at the Provost's Seminar, MIT.

Rota shares Husserl's belief that a new Galilean revolution will come about to create an alternative, soft mathematics, that will establish theoretical laws through idealizations that run counter to "common sense."

Platonic (Deistic) Philosophies of Mathematics vs. Humanistic Philosophies

The traditional platonic philosophy of mathematics, with its emphasis on the single question "why is mathematics true?" and its espousal of various "myths," is suffering from "dead-end-itis." But there will be recitation of these myths before we can safely recite the *Nunc Dimittis*.

The mythic unity of mathematics was already threatened in Poincaré's day by the sheer size of the material available. The riches of mathematics, without contemplative judgments, would, in the words of Poincaré "soon become an encumbrance and their increase will produce an accumulation as incomprehensible as all the unknown truths are to those who are ignorant." The unity is now further threatened by self-contained, self-publishing chat groups.

The classic Euclidean mode of exposition and teaching: "definition, theorem, proof" has been seriously questioned as not providing a realistic description of how mathematics is grasped, utilized or created.

Here are quotes from four distinguished mathematicians that bear on this. From Brian Rotman:

> By giving mathematicians access to results they would never have achieved on their own, computers call into question the idea of a transcendental mathematical realm. They make it harder and harder to insist, as the Platonists do, that the heavenly content of mathematics is somehow divorced from the earthbound methods by which mathematicians investigate it. I would argue that the earthbound realm of mathematics is the only one there is. And if that is the case, mathematicians will have to change the way they think about what they do. They will have to change the way they justify it, formulate it and do it.
> –"The Truth about Counting," *The Sciences*, Nov. 1977

From Richard Hamming:

> I know that the great Hilbert said "We will not be driven out of the paradise that Cantor has created for us." And I reply: "I see no need for walking in." - "Mathematics on a Distant Planet," *American Mathematical Monthly*, v.105, 1998.

From Solomon Feferman:

> I think the Platonistic philosophy of mathematics that is currently claimed to justify set theory and mathematics more generally is thoroughly unsatisfactory and that some other philosophy grounded in inter-subjective *human* conceptions will have to be sought to explain the apparent objectivity of mathematics.

From Gregory J. Chaitin:

> In the end it wasn't Gödel, it wasn't Turing, and it wasn't my results that are making mathematics go in an experimental direction. The reason that mathematicians are changing their habits is the computer. - *The Limits of Mathematics*

A Personal Illumination

Here, then, are some of the "tensions of mathematical texture" that I perceive. Today's scientist/mathematician spends his or her days in a way that is vastly different from fifty or even twenty years ago. Thinking now is accomplished differently. Science is undergoing a fundamental change; it may suffer in some respects, but it will certainly create its own brave new world and proclaim new idealisms.

I think there will be a expansion of what has traditionally been considered to be valid mathematics. In the wake of this, the field will again be split just as it was in the late 1700s when it split into the pure and the applied. As a

Phil Davis, 1972.

consequence, there will be the "true believers" pursuing the subject pretty much in the traditional manner, and the "radical wave" pursuing it in ways that will raise the eyebrows and the hackles of those who will cry "they are traitors to the great traditions."

In his autobiography, Elias Canetti, Nobelist in literature (1981), speaks of an illumination he had as a young man. Walking along the streets of Vienna, he saw in a flash that history could be explained by the tension between the individual and the masses.

Walking the streets of my home town, I got an illumination: the history of future mathematics will be seen as the increased tension and increased interfusion, sometimes productive, sometimes counterproductive, between the real and the virtual. But should I call this an illumination when its signs can be read everywhere, when the changes from the present to the future seems to have the inevitability of a locomotive thundering down the tracks?

How these elements will play out is now a most excellent subject for writers of mathematical fantasies.

• • •

A note to Mogens Niss, who some years ago, asked me to tell him how I came to my views.

Dear Mogens,

The British historian R. G. Collingwood wrote that he learned from watching his parents who were artists that a painting is never finished: it is abandoned.

So too with this book. Nonethless, I hope that the outlines of my education are now sufficiently clear to you.

P.J.D.

Acknowledgements

I wish to acknowledge the help, encouragement, and enlightenment I received from the following individuals and institutions.

Robert Barnhill; Fred Bisshopp; I. Edward Block; Christa Binder; Bernhelm Booss-Bavnbek; Sussi Booss-Bavnbek; Rosemary Chang; Gail Corbett; Constantine Dafermos; Susan Danforth; Ernest S. Davis; Hadassah F. Davis; Joseph M. Davis; Marguerite Dorian; Paul Ernest; John Ewing; Michael O. Finkelstein; Stuart Gehman; John Guckenheimer; Pete and Ruth Haynsworth; Reuben Hersh; Arieh Iserles; Ariel Jaffee; Ann Kostant; Laura Leddy; Kathryn Maier; David Mumford; Igor Najfeld; Mogens Niss; Clara Park; David Park; Glen Pate; Alice and Klaus Peters; George Phillips; David Pingree; Brian Rotman; Peter Schmitt; Anne Spalter; Trevor Stuart; James Tattersall; Andries and Debbie Van Dam; Jerome Weiner; Margaret Wertheim; Alvin White; Roselyn Winterbottom.

Also: Division of Applied Mathematics, Brown University; IMFUFA, Roskilde University Centre, Denmark; Urania Theatre, Berlin.

The SIAM NEWS (Society for Industrial and Applied Mathematics) and the Humanistic Mathematics Network Journal have generously allowed me to use a few portions of my articles published in those periodicals.

Two people require special citations. Bernhelm Booss-Bavnbek was initially a co-author in an early version of this book. For almost a decade I have profited from our conversations, e-mail dialogues, his numerous faxed writeups, and his generous consent to my using some of his material and ideas.

My wife, Hadassah F. Davis, has been the absolute without-whom-not of this book, and I have often said to her and continue to say "thy word is a lamp unto my feet."

Books by Philip J. Davis

Mathematical Monographs

Interpolation and Approximation
Methods of Numerical Integration (with Philip Rabinowitz)
Circulant Matrices
Spirals: From Theodorus to Chaos

Texts

The Mathematics of Matrices

Philosophy

The Mathematical Experience (with Reuben Hersh)
Descartes' Dream (with Reuben Hersh)
No Way: The Nature of the Impossible (with David Park)

Recreational

The Lore of Large Numbers
3.1416 and All That (with William Chinn)

Fiction (Possibly: Faction)

The Thread
Thomas Gray, Philosopher Cat
Thomas Gray in Copenhagen

Biography

Mathematical Encounters of the Second Kind